MUSINGS OF THE MASTERS

An Anthology of Mathematical Reflections

To the memory of my parents,
George and Nazeera,
and my siblings
Anne, Frank, and Lucien

I should like to express my deep gratitude and appreciation to Prof. Gerald Alexanderson. His outstanding scholarship, his patience, and his friendly persuasion have been vital in bringing this work to fruition.

ISBN: 0-88385-549-6
Library of Congress Catalog Card Number: 2004104239

Current Printing (last digit):
10 9 8 7 6 5 4 3 2 1

MUSINGS OF THE MASTERS

An Anthology of Mathematical Reflections

Edited by
Raymond G. Ayoub

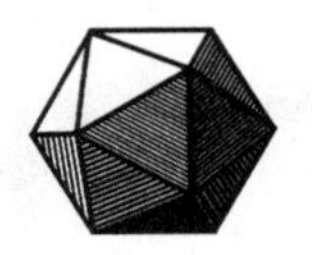

Published and Distributed by
The Mathematical Association of America

The Spectrum Series of the Mathematical Association of America was so named to reflect its purpose: to publish a broad range of books including biographies, accessible expositions of old or new mathematical ideas, reprints and revisions of excellent out-of-print books, popular works, and other monographs of high interest that will appeal to a broad range of readers, including students and teachers of mathematics, mathematical amateurs, and researchers.

777 Mathematical Conversation Starters, by John dePillis

All the Math That's Fit to Print, by Keith Devlin

Carl Friedrich Gauss: Titan of Science, by G. Waldo Dunnington, with additional material by Jeremy Gray and Fritz-Egbert Dohse

The Changing Space of Geometry, edited by Chris Pritchard

Circles: A Mathematical View, by Dan Pedoe

Complex Numbers and Geometry, by Liang-shin Hahn

Cryptology, by Albrecht Beutelspacher

Five Hundred Mathematical Challenges, Edward J. Barbeau, Murray S. Klamkin, and William O. J. Moser

From Zero to Infinity, by Constance Reid

The Golden Section, by Hans Walser. Translated from the original German by Peter Hilton, with the assistance of Jean Pedersen.

I Want to Be a Mathematician, by Paul R. Halmos

Journey into Geometries, by Marta Sved

JULIA: a life in mathematics, by Constance Reid

The Lighter Side of Mathematics: Proceedings of the Eugène Strens Memorial Conference on Recreational Mathematics & Its History, edited by Richard K. Guy and Robert E. Woodrow

Lure of the Integers, by Joe Roberts

Magic Tricks, Card Shuffling, and Dynamic Computer Memories: The Mathematics of the Perfect Shuffle, by S. Brent Morris

The Math Chat Book, by Frank Morgan

Mathematical Adventures for Students and Amateurs, edited by David Hayes and Tatiana Shubin. With the assistance of Gerald L. Alexanderson and Peter Ross

Mathematical Apocrypha, by Steven G. Krantz

Mathematical Carnival, by Martin Gardner

Mathematical Circles Vol I: In Mathematical Circles Quadrants I, II, III, IV, by Howard W. Eves

Mathematical Circles Vol II: Mathematical Circles Revisited and *Mathematical Circles Squared,* by Howard W. Eves

Mathematical Circles Vol III: Mathematical Circles Adieu and *Return to Mathematical Circles,* by Howard W. Eves

Mathematical Circus, by Martin Gardner

Mathematical Cranks, by Underwood Dudley

Mathematical Evolutions, edited by Abe Shenitzer and John Stillwell

Mathematical Fallacies, Flaws, and Flimflam, by Edward J. Barbeau

Mathematical Magic Show, by Martin Gardner

Mathematical Reminiscences, by Howard Eves

Mathematical Treks: From Surreal Numbers to Magic Circles, by Ivars Peterson

Mathematics: Queen and Servant of Science, by E.T. Bell

Memorabilia Mathematica, by Robert Edouard Moritz

Musings of the Masters: An Anthology of Mathematical Reflections, edited by Raymond G. Ayoub

New Mathematical Diversions, by Martin Gardner

Non-Euclidean Geometry, by H. S. M. Coxeter

Numerical Methods That Work, by Forman Acton

Numerology or What Pythagoras Wrought, by Underwood Dudley

Out of the Mouths of Mathematicians, by Rosemary Schmalz

Penrose Tiles to Trapdoor Ciphers ... and the Return of Dr. Matrix, by Martin Gardner

Polyominoes, by George Martin

Power Play, by Edward J. Barbeau

The Random Walks of George Pólya, by Gerald L. Alexanderson

Remarkable Mathematicians, from Euler to von Neumann, by Ioan James

The Search for E.T. Bell, also known as John Taine, by Constance Reid

Shaping Space, edited by Marjorie Senechal and George Fleck

Sherlock Holmes in Babylon and Other Tales of Mathematical History, edited by Marlow Anderson, Victor Katz, and Robin Wilson

Student Research Projects in Calculus, by Marcus Cohen, Arthur Knoebel, Edward D. Gaughan, Douglas S. Kurtz, and David Pengelley

Symmetry, by Hans Walser. Translated from the original German by Peter Hilton, with the assistance of Jean Pedersen.

The Trisectors, by Underwood Dudley

Twenty Years Before the Blackboard, by Michael Stueben with Diane Sandford

The Words of Mathematics, by Steven Schwartzman

MAA Service Center
P.O. Box 91112
Washington, DC 20090-1112
800-331-1622 FAX 301-206-9789

Introduction

This anthology is a collection of articles written by renowned mathematicians of the past century. The articles are on a variety of topics that, for want of a better name, shall be referred to as "humanistic." An important criterion, thereby limiting the choice, is that the articles should be accessible to the general reader who need not have a technical knowledge of mathematics. These stipulations do not certify that the articles are easily understood! Nevertheless, it is hoped that the reader will find in them much of interest.

A motive in collecting these articles stems, in part, from a curiosity concerning the creative process in mathematics together with a curiosity concerning the essence of this remarkable subject. While some articles deal with this theme, it should quickly be said that these essays do not cast any decisive illumination on these matters!

Some of the essays have a common thread, while several others deal with a specific topic: Maak's essay on Goethe and mathematics, and Severi's article on Leonardo.

For the benefit of the non-specialist a few general remarks may be useful in placing the discipline of mathematics in a quasi-historical setting.

Mathematics has occupied a central position in the history of civilization. This is, in part, because of its antiquity, its universality, its utility, and its enigmatic dual nature. It spans divers cultures; on the one hand its dual nature allows it to function in a very wide spectrum of applications, and on the other hand in the "purest" of the arts and humanities. Over the years it has been revered by some and despised by others; it has been extolled by its advocates and condemned by its opponents. (St. Augustine, for example, was convinced that mathematicians were in league with the devil, but some scholars think that he conflated mathematics with astrology!) Such a discipline invites some comments.

Much has been written about the history of the subject and it is not necessary to review this in any detail. It is very likely that mathematics originated in the earliest civilizations as an empirical discipline designed to meet the practical needs of an organized society. It is equally likely that, given the inherent intellectual curiosity of some humans, it was not long thereafter that this innate curiosity stimulated a study of mathematics as a purely intellectual discipline. This natural inclination could account for the very remarkable reality that some mathematical facts were discovered, apparently independently from one another, in areas and cultures widely separated in space and in time. A notable example is the result known as the "Pythagorean Theorem" which was discovered independently not only by the ancient Greeks but by scholars on the Indian sub-continent as well as mathematicians in China.

Like other arts and sciences the progression of mathematics has not been continuous but has waxed and waned and the effect of cultural influences upon this variation has attracted the attention of some writers. There are periods of feverish creativity and fallow ones. As a side remark, it should be stressed that contemporary mathematics is experiencing a feverish, if not hectic, period of creativity.

Nor is it necessary to dwell on these matters, except to remind readers that about 600 B.C. in the hands of the Greeks, in what some have called the "golden age," mathematics gradually became formalized. With occasional lapses, this formalization has continued to this day. As a result, mathematics has acquired a dual nature—that of a purely intellectual pursuit and that of a mechanism for coping with and modeling problems of the "real" world. Both of these facets have contributed to the intellectual history and the technological progress of mankind. For want of better descriptive terms, the theoretical pursuit will be referred to as the 'sacred' mode and the functional or practical side shall be called the 'secular' mode. Consumers of mathematics should not take offense at this choice of words. It is a fruitless pursuit to debate the relative merits of these modes since both have played, and continue to play, a significant role in societies—each has acted as a catalyst to the other.

In its sacred mode, the principal activities and principal objectives of mathematics can be described as the study of structures and the discernment of patterns. Although somewhat vague it would appear that this is as close as one can get to a focus or purpose of mathematical activity. In reality, however mathematics has apparently no well-defined objective. At any rate there has been no articulated objective upon which mathematicians have agreed. Different practitioners pursue their favorite structures with enthusiasm and sometimes recklessness! This is not to say that the subject lacks harmony and cohesion—it is a miraculous fact that in some mystical way, many of the various parts seem, eventually, to fit together into an organic whole. It should be said that the detection and discernment of this cohesion poses a constant challenge, especially to those who are fearful that the subject may disintegrate into an anarchy of mathematical fiefdoms.

A mathematician, primarily in the sacred mode, is more or less free to define a set of axioms governing a structure and then draw inferences as to its properties. The mathematician is often free to search for a mathematical object with prescribed properties, however contrived these may seem to be. In reality there are subtle forces, not clearly understood, which prevent chaos from setting in.

Although there is a certain randomness in the choice of structure or property which is chosen for study, the history of mathematics shows that there is a ruthless selectivity that takes place, and in a manner that is far from being understood. The discipline casts off those parts that do not give it its proper nourishment. This selectivity is made, not by the practitioners but in a curious way, by the subject itself—the whole mathematical organism deprives the mathematician of his or her autonomous role, retaining only those parts that it deems essential. A hazardous activity then, is to try to predict the fate of a mathematical structure or object and one of the nightmares of a mathematician is to contemplate the future fate of his or her work! There is the story, possibly apocryphal, of Bertrand Russell's dream in which he saw a phantom walking through the stacks of a library with a roaring fire below. The phantom examines a book taken from the shelf and either returns it to the shelf or casts it into the flames. Coming upon the *Principia*, a monumental work of Russell and Whitehead, the phantom pauses, hesitates for a moment—at which time Russell awoke from his dream!

In the secular mode, the mathematician or scientist when attempting to model a "real world" problem is constrained by the exigencies of the problem and cannot stray too far from the properties being modeled. To be sure, simplifying assumptions may, and often must, be made to render the problem accessible to whatever mathematical techniques are available. It has been repeatedly acknowledged that mathematics has been extraordinarily effective in modeling the real world. Structures or mathematical constructs that have been studied as intellectual pursuits become, possibly after many decades, mental images of real world phenomena. The source and reason for this phenomenon is shrouded in mystery and has been the focus of much speculation but no satisfactory explanation has been forthcoming. Mathematics appears to permit a practitioner to conduct thought experiments, using different structures and different relations until some harmony with observation is achieved. It would appear as though in some unaccountable sense the mental constructs have mirrored the real world.

Although the distinction between secular and sacred is not sharply defined, it is in the sacred mode that mathematics most decisively exhibits its humanistic side. It is in this aspect of the subject that mathematicians, over the centuries, have stressed its aesthetic qualities that, on the face of it, appear paradoxical. For many it is an aesthetic experience perceived not through the senses, as one would normally expect for an aesthetic experience, but through the realm of thought. It is an art evoking powerful emotions but without the intervention of the senses.

Before continuing this train of thought, it is worthwhile pausing to make a distinction, again not too well defined. As in the arts, a distinction is made, however often blurred, between the popular and the serious. In the graphic arts for example, distinction is often made between so-called "commercial" and other forms of art and in the realm of poetry, for example, limericks are separated from serious poetry. To be sure some of these examples can become elevated to more significant levels. So in mathematics, there is a lighter side such as number puzzles and magic squares. Again some of these have found their way into the classics.

A few examples of mathematicians extolling the beauty of the subject can be noted. Writing to Sophie Germain, Gauss expressed himself as follows: "The enchanting charms of this sublime subject do not reveal themselves except to those who have the courage to plumb its depths." Poincaré wrote: "We are astonished to find the senses involved in mathematical proofs which should appeal only to the intellect. There is...a sense...of beauty of the harmony of number and form, of geometric elegance. It is a genuine aesthetic sense which all mathematicians know." From the pen of Russell we read "mathematics rightly understood, possesses not only truth but supreme beauty." These examples could be multiplied tenfold but we end with an anonymous quote: "When we read the memoirs of Gauss ... does not the overall structure recall one of the marvelous temples ... raised to the Hellenic divinities."

In these and other cases, the beauty and harmony are palpable but not through the ordinary senses.

Another question that arises is that of the reality of mathematics—where does it reside? The issue has provoked much discussion and debate, often pitting one side against another; its ultimate resolution is far from certain. Indeed, the point of view one takes is as much a matter of quasi-theological belief as it is a matter of intellectual certitude.

Some hold that its reality resides entirely internally, that is, that mathematics is created as a musician might create a quartet or an artist paint a Madonna. Others say that it is entirely external—that mathematics is a process of discovery. Not surprisingly others take a middle of the road view, that it is a combination of the two. A well-known proponent of this last view is Kronecker who in his famous dictum wrote: "God made the integers—all else is man made." G.H. Hardy is an advocate of the second view: "We believe," he wrote, "that mathematical reality lies outside of us, that our function is to discover or observe it and the theorems which we grandiloquently describe as our own creations, are simply our notes of our observations." Hardy's view is supported by Hermite who wrote: "There exists, if I am not mistaken, a whole world which is the set of all mathematical truths to which we have access by intelligence, as there exists a world of reality—both independent of us and both divine creations." Hilbert represents the other extreme writing "mathematics is a meaningless formal game." And the debate continues.

To be sure the discipline "philosophy of mathematics," a time-honored pursuit, has flourished in recent years. It is difficult, however, for one uninitiated, to find enlightenment here. The reason is that much of the discourse is phrased in technical philosophical vocabulary—terms which are remote from the layperson's experience. Such terms as "Platonism," "Cartesian duality," "epistemology," Hegelianism, post modernism, existentialism, and so on, are alien to most laypersons' experiences.

In view of its elusive and recondite nature, one might expect that mathematicians would rise to the challenge and write more extensively about their subject. This is, surprisingly, not the case. With some notable historical exceptions, such as Descartes, Leibniz, and in more recent decades Brouwer, Poincaré, Hadamard, Hardy, and Weyl, mathematicians have been curiously reluctant to write on matters that lie outside the strict confines of mathematics proper and mathematical research. It is not easy to account for this reluctance but some contributing factors suggest themselves.

First, writing about mathematics and areas contiguous to it, is a somewhat daunting prospect, for it implies that we know what we are talking about. As noted above the nature of the subject is elusive, and this elusiveness has been widely recognized by mathematicians. There have been many attempts to define mathematics and these are so unsatisfying that this writer's favorite is that of Bacon who wrote that "mathematics is the study that makes men subtile (sic)"! This is amusing but hardly enlightening.

Second, mathematicians by their training and practice have become highly immersed in habits of precise thought and reasoning in their discipline. To write about other subjects—ethics, religion, politics—is a dismaying prospect, for these subjects are by their nature imprecise and a discourse requires a transition into different modes of thought, a transition that is not easy to make. Many mathematicians have, nevertheless, ventured very successfully into areas such as philosophy, and fiction.

There is another factor that influences this inhibition and reticence. This has to do with the culture of the mathematical community. The traditions of the past were somewhat rigid and these same traditions viewed the writing of textbooks or popular works on mathematics with a slight degree of scorn. Such writing was well received after the writer had established a reputation as a research scholar. Many examples may be cited.

As a result of these and other induced inhibitions, society has been largely deprived of the musings of first-rate mathematical minds—a deprivation that is unfortunate. Some few courageous souls however, possibly coming initially from different traditions, have set pen to paper and have left a worthwhile legacy. To be sure a gift in one area of the intellect does not necessarily carry over to other areas—how else can we account for the outrageous excesses of which some mathematicians have been guilty. Fortunately however, there has been a moderate amount of transfer and the fruits of this transfer are articles by prominent mathematicians writing on topics sometimes contiguous to, and sometimes remote from, their principal occupation—that of mathematical research.

These articles have been assembled and are offered in a single place. They span roughly a century in time and a wide range in subject. They are by mathematicians acknowledged by their peers as outstanding creators whose work has added richly to the discipline. Each article is preceded by a brief biographical sketch of the author and a short indication of the content. Many of them elaborate on some of the issues that have been raised above. To be sure in many articles there is a common theme—the attempt to define the role of intuition in mathematical activities— intuition as contrasted with pure formalism, a contrast that can be briefly, if too simply, described by asking whether mathematics is invented or discovered.

We hope that the appearance of these articles in a single place will encourage mathematicians to share with others their perceptions on this beautiful subject, or on other matters dealing more directly with the humanistic side of our nature.

Contents

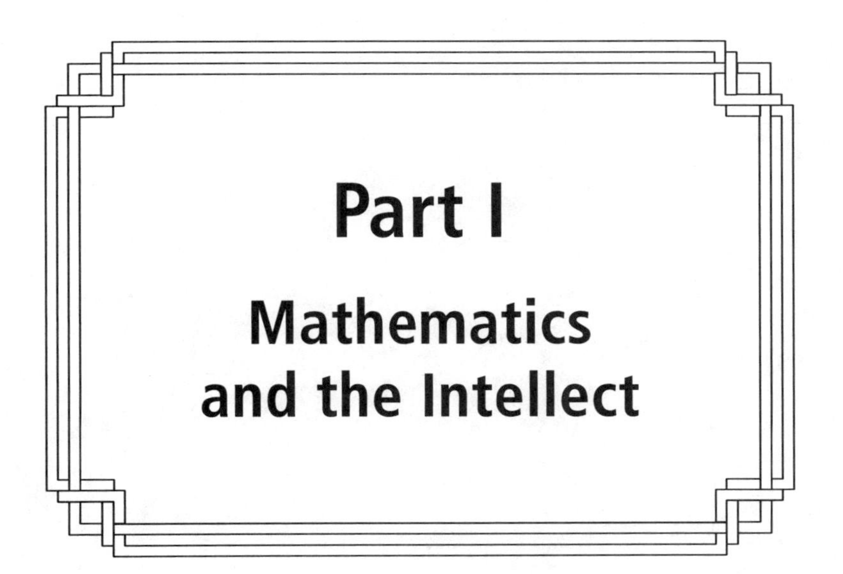

Part I

Mathematics and the Intellect

Mathematics is like checkers—it is suitable for the young, not too difficult, amusing, and without peril to the state. —Plato (ca. 429–347 B.C.)

Mathematics and Thinking Mathematically

Mary Lucy Cartwright

Dame Mary Lucy Cartwright (1900–1998), D.B.E., F.R.S., to give her full name and honors, was born in Northamptonshire, the daughter of a rector. She was educated at home and subsequently at a "public" school in Salisbury. At this time the family was struggling to cope with the tragic death of Cartwright's two older brothers in World War I. She entered St. Hughes College, Oxford, in 1919 and, although she was engrossed in the study of history, she opted to "read" mathematics. She was one of only five women studying mathematics in Oxford at that time. Having devoted more time to history than to mathematics at school, the program in mathematics was a challenge but her natural mathematical endowments enabled her to earn a first class degree in 1923.

Not wishing to put a strain on the family finances, she taught school for four years and then returned to Oxford where she was awarded a D.Phil. in 1930 under the supervision of G. H. Hardy and E. C. Titchmarsh. In 1934 she was appointed to a faculty position at Girton College (a women's college in Cambridge University) where she was director of studies in mathematics.

Her mathematical research has ranged over a wide field of classical analysis and has been seminal in many areas, especially in integral equations.

In 1939 the Department of Scientific and Industrial Research asked Cartwright to help in solving "certain very objectionable looking equations occurring in connection with radar." In this enterprise, she collaborated with J. E. Littlewood who (tongue in cheek) described her as "the only woman in my life to whom I have written twice in a single day." He held her in high esteem as a colleague.

In 1949 she became mistress of Girton, a position that placed great demands upon her time, demands she cheerfully and willingly gave. It is said that she provided quiet, unassuming and clear-headed leadership of the college during a time of many challenges. Although shy by nature, she managed to interview all incoming students and to regale her younger faculty with tales of her foreign adventures.

Her earlier interest in history never left her entirely, for her work is permeated with historical perspectives that add interest and dimension to her work.

Her scientific work was recognized by election to the Royal Society in 1947. She also served as president of the London Mathematical Society from 1961 to 1963 as well as President of the Mathematical Association. She received honorary degrees from Cambridge, Edinburgh, Leeds, Hull, Wales, and Oxford. In addition, she was awarded the Sylvester Medal of the Royal Society in 1964 and the De Morgan Medal of the London Mathematical Society in 1968. In June 1969 she was made a Dame Commander of the British Empire for services to mathematics.

She died in 1998 and is revered for, among other things, having helped pave the way for more widespread recognition of women in mathematics.

Editor's Preface

In this essay, Mary Cartwright endeavors to throw some light on the difficult question of what constitutes "mathematical thinking," especially abstract thinking. The power to engage in complete abstraction, she notes, comes very slowly and, in the case of most individuals, comes not at all. Yet as we note below, surprising levels of understanding abstract ideas can be achieved. At present the mechanism of that achievement appears to be an impenetrable phenomenon of human thought. Present and future research in the neurological or psychological nature and basis of abstract thought may cast some light on this currently intractable problem.

Although abstract thinking arises in numerous contexts, its most conspicuous manifestation seems to be in mathematics. In the ongoing debate involving "nature" and "nurture," that is, whether a characteristic is genetically or culturally acquired, we suggest that abstract thinking is a *learned* attribute and in the case of mathematics, it seems to be intimately intertwined with the specific context.

Mary Cartwright points to another distinctive characteristic of mathematics, viz. that consumers of mathematics (engineers, physicists, etc.) invariably relate the mathematical construct and content to the individual physical phenomenon that interests them and that governs the phenomenon. This same process takes place at different levels of abstraction. It is said that Newton, for example, invariably associated differential equations with physical events. In a somewhat different direction, it is reported that a noted algebraist of the early part of the 20th century always thought of a linear transformation as a *matrix*. For early group theorists, a group was generally thought of as a set of *permutations*. In neither of these cases was the current abstract form used as a basis for deeper investigations. We hasten to add, however, that the less abstract point of view often led, nevertheless, to some deep and enduring discoveries.

To achieve a higher level of abstraction, it is often the case that a mathematical construct is disassembled into its component parts and reassembled, more nearly "to the heart's desire." The process may lead to a higher level of abstraction and, according to Cartwright, could thus lead to a greater understanding.

Returning to the matter of abstraction, it is a very interesting, curious and challenging question to account for the fact that in the historic evolution of mathematics, higher levels of abstraction are achieved and better understood, apparently quite simply *with the passage of time*. Each new generation is better able to cope with greater levels of abstraction. Since genetic differences do not occur in such short periods of time, we must account for this phenomenon by a change in the environment—teachers and writers of mathematics are better able to work with abstraction and are better able to articulate it. The passage of time has miraculously enhanced the clarity. Obviously this has been a consequence of evolving cultural changes. One of the most notable examples is the evolution of the Hindu-Arabic system of enumeration and the computational algorithms associated with it. Beginning from an arcane and almost mystical study in the 9th till the 12th century, to a system of computational algorithms that were, at first, limited in use to specially trained calculators in the 15th and later centuries, the system is now routinely taught to children in grade schools. It would seem that each generation contributed to a greater clarity and systematization of the underlying abstraction. It could now more easily be communicated at an early stage in a child's development.

Much has been written in recent years as people endeavor to understand the process of learning mathematics among children and to decipher the abstractions that underlie it. It is assumed that such investigations will cast light on the mathematical way of thinking. Among the leading investigators have been Piaget and Inhelder. They have unearthed many very interesting phenomena related to mathematical learning among children. However, as many elementary school teachers will testify, the enterprise has not been as effective as was originally hoped for and the most recent researches have not satisfactorily resolved the basic problems confronting teachers of mathematics, that is, the learner's ability to cope with abstraction.

Professor Cartwright however, has made some thoughtful observations.

The lecture was given as the Samuel Newton Taylor Lecture at Goucher College, January 30, 1969.

This year I find myself in the Division of Applied Mathematics at Brown University. Not every University or College has a separate department of applied mathematics, but for over thirty years I was classed as a pure mathematician in the University of Cambridge, England, and some applied mathematicians there prefer to call themselves theoretical physicists. Moreover, I once heard a geophysicist with a mainly mathematical training say that he used 'applied mathematics' as a term of abuse, meaning stuff which was not good mathematics and not really relevant to any physical problem. All these factors have made me think about the borderline between mathematics and its applications, not only to physical problems of a more or less traditional type, but also to statistical, economic, and industrial problems.

It is well known that the origins of some of the most abstract pure mathematics can be traced through the theory of Fourier series to a problem about vibrating strings, or through the theory of irrational numbers to Greek geometry and Egyptian devices for measurement of right angles, but the pure mathematicians of the last 100 or 150 years have been pursuing the mathematics for its own sake without any thought of vibrating strings. On the other hand many major new developments in pure mathematics were initiated quite specifically for the purpose of using them in some application. For instance this is certainly true of Newton's contributions to the calculus, and of probability theory, and this still seems to be happening in operations research and control theory. In distinguishing pure mathematics from applied two questions seem to arise. Is the work truly abstract and separated from all applications? And is it any more mathematical if it is truly abstract and pursued strictly for its own sake?

If we delve into the beginnings of mathematical thought in very young children or primitive peoples, there is plenty of evidence to show that the power of complete abstraction comes very slowly, and indeed to many people it probably only ever comes in a very restricted sense. A number of eminent people take the view that thought begins with the idea of actions performed in the mind only, that is to say operations. According to Piaget an ordinary child, by the time the child is two, can work out how he is going to do something before he does it, *provided* that the situation is simple and is familiar to him, but in order to understand abstract mathematical concepts such as 1, 2, 3, 4,…, the child has to move from perceptions arising from his environment and actions to the abstractions, a long and gradual process. Much work has been done by Piaget and Innhelder on the child's conception of space, and, for instance, its powers to distinguish between different kinds of figures such as a circle, a square, and a circle with a little one either inside or outside. Their experiments have thrown much light on the development of numerical, spatial, and physical concepts of a very elementary kind among young children, but it seems doubtful to me whether the abilities tested are always truly mathematical. For young blackbirds will gape at a piece of black cardboard consisting of one large circle and two small ones attached to it, but they only gape at the small circle whose size is a certain proportion of that of the large circle. This indicates that the ability to distinguish between certain shapes may have psychological foundations.

H. and H. A. Frankfort in an essay on myth and reality point out that ancient man could reason and work out the causes of things, but worked on very different hypotheses from ours. The primitive mind asks 'who' when it looks for a cause, and cannot withdraw far from perceptual reality. When the river does not rise, the river has *refused* to rise, and so the river or the gods intend to convey something to the people. At the same time primitive man used symbols much as we do, but he can no more conceive them as signifying, yet separate from, the gods or powers than he can consider a relationship—such as resemblance—as connecting, and yet separate from, the objects compared. Hence there is a coalescence of the symbol and what it signifies, as there is coalescence of two objects compared so that one may stand for the other.

Frankfort then gives an example of this coalescence in which pottery bowls with the names of hostile tribes were solemnly smashed at a ritual by the Egyptians in the belief that real harm was done to the enemies by the destruction of their names. It may seem a far cry from this to modern mathematics, but Bochner has drawn a parallel between mathematics and myth, and replaced myth by mathematics in some of Frankfort's sentences. I am not prepared to go as far as he does by replacing the word myth by mathematics in a sentence which then asserts that mathematics transcends reasoning in that it wants to bring about the truth it proclaims. However, in the ritual we have two fundamental features of mathematics, symbols representing something and operations on those symbols representing operations on the thing itself. Symbols and notation are part of the essential basis of mathematics, and I believe that the development and standardization of a good notation is an extremely important part of the development of mathematics.

If we turn to the extreme other end of the scale, we run into another kind of difficulty in separating the mathematics from its applications. Some pure mathematicians seem to do their mathematical thinking in terms of idealized physical and spatial ideas. The late G. H. Hardy, who taught me, was very much against applied mathematics, but in a footnote to a joint paper with J. E. Littlewood published in a Swedish periodical he wrote that a certain problem is most easily grasped in terms of cricket averages. Norbert Wiener would translate a mathematical problem into the language of Brownian motion, and I believe that his thinking was completely abstract although I do not know the theory, or remember what he said well enough to be quite sure. Hadamard has described his visualization of the proof that there is a prime greater than 11. To consider all prime numbers from 2 to 11, i.e., 2, 3, 5, 7, 11 he visualized a confused mass. Forming the product $2 \times 3 \times 7 \times 11 = N$, since N is large, he visualized a point remote from the mass. Increasing the product by 1 he saw another point a little beyond the first. $N + 1$, if not a prime, is divisible by a prime greater than 11; Hadamard saw a place between the mass and the first point. This seems to me to be a sort of mathematical shorthand and would certainly have to be translated back to numbers before it could be communicated to anyone else.

As I said earlier, I have until now always been classed as a pure mathematician, but Professor J. E. Littlewood and I did a lot of work on the theory of ordinary differential equations arising from problems of radio engineering. Littlewood is also a very pure mathematician in many ways, but he worked on antiaircraft gun fire in the First World War, and he translated our problems, which were suggested by radio values and oscillations, capacitance and inductance, etc., into dynamical problems and called all the solutions of our equations 'trajectories' as if they were the paths of missiles shot from a gun. In the radio problems there are oscillations with negative damping, and so we had periodic trajectories going up and down over and over again, and I am sure that the abstraction was complete although there was often a certain woolliness until the argument was complete, just as in Hadamard's visualization. Between these two extremes there are sonic users of complicated mathematics, physicists and engineers in particular, who are thinking all, or nearly all, the time in terms of the physics of the problem. Engineers have consulted me about a number of different types of problem, radio, control theory, oscillations of stretched wires; they usually come with some equations and very little explanation. I have to ask a lot of questions before they tell me everything relevant to the mathematical problem. It seems difficult for them to think in abstract mathematical terms, the symbols to them seem to mean the engineering concepts, currents and circuit constants such as impedance and inductance. This is important in two ways. The engineers have mental reservations and can check at every stage because they visualize how the physical system works. On the other hand they find it difficult to apply the mathematical processes used in one field to any other physical problem, even if they are just as relevant there. Some years ago at a conference for engineers I was asked to speak on Liapunov's method for stability problems. I described the basic principles as simply as I could, and after I spoke Professor Parks lectured on applications of the method. Many in the audience commented that the order of our lectures should have been reversed, and that they would have understood my lecture much better if they had understood that I was talking about the phase plane. It is possible that this was partly a question of notation and terminology, but I believe that they could do advanced mathematics best by thinking of it in terms of their particular engineering problems. The Liapunov method was developed mainly in connection with control engineering and by now has adopted much of its terminology, but the mathematics arising there need to be abstracted and put in a form which makes it available in connection with other applications. Problems of ordinary differential equations have arisen in connection with astronomy, ballistics, radio engineering, control theory, mechanical oscillations of machinery; each application has special features, and the theory of it was often developed in a correct logical form quite a long way before it was fitted into the general theory of ordinary differential equations as pure mathematics. The individual who formulated the equation and asked the question is, in the sense of my title, thinking mathematically, but he is not doing mathematics until he operates on his symbols. Please note that I do not say 'asked for a solution of the equation' because, although he may say that, he really wants to know something about the solutions in general. Is there a periodic solution? Is it stable? Will it remain stable if I change a certain parameter? Will the period be longer or shorter? He may find the methods which he needs in the literature and do the work himself. He may find a mathematician to help him. Although I myself have helped to develop the general theory and settle certain theoretical problems, I do not think that I have ever produced a result useful for any specific practical problem when it was needed. For soon after Littlewood and I began work on these problems, it was realized that the variations in individual thermionic valves was so great that precise mathematical results were not worth the trouble, and satisfactory experimental determinations could be more easily obtained. In recent times the person who formulates the mathematical statement of a physical or other real life problem usually does not do anything very original in the mathematical handling of it, although some interesting purely mathematical work on matrices appears in journals concerned with computing or applications to economics, detached from other pure mathematics.

To sum up so far I believe that the dividing line between strictly abstract thinking in mathematics and thinking in terms of the real world is by no means clearly defined and some of the major developments in mathematics such as the calculus were thought out more or less in terms of the real world. Further abstraction does not necessarily make the mathematics any better. For the Babylonian schoolmasters constructed sets of most complicated artificial formulae, perhaps 200 on one tablet, for their pupils to simplify. Their mathematics was sufficiently abstract for them to be indifferent whether they added the number of men to the number of days. In present circumstances this seems abstraction at its worst, but perhaps then it was a step forward. The Babylonians must have developed the laws of arithmetic a long way to set these complicated exercises, but mainly for practical purposes whether it was accounting or astronomy.

Now let us turn to those who do mathematics for its own sake. I should like to begin with the Hindu who in about 1200 B.C. wrote, "As crests on the heads of peacocks, as the gems on the hoods of snakes, so is ganita, mathematics, at the top of the sciences known as the Vedanga." Ganita is literally the science of calculation and in the early days it consisted of finger arithmetic, mental arithmetic, and higher arithmetic in general. At first it included astronomy, but geometry belonged elsewhere. At one stage higher mathematics was called 'dust work' because it was done in sand spread on the board or on the ground. We owe our so-called Arabic numerals to the Hindus, and they advanced a long way in algebra very early.

Most people consider that the Greeks were the first to do mathematics for its own sake and to realize the need for proof. The word 'mathema' meant originally a subject of instruction, but very early it was restricted to mathematical subjects among which Pythagoras included geometry, theory of numbers, sphaeric (or spherical trigonometry used for astronomy), and music. They classified numbers not only as odd and even, but as even-even, 2^m; even-odd, $2(2n+1)$; odd-even $2^{m+1}(2n+1)$, and also proved that there are an infinity of primes. I doubt whether they could calculate as well as the Babylonians, but probably that did not attract them, and also they lacked the incentives provided by the government of a far flung empire. I feel that I have to remind myself of the difficulties due to the absence of convenient symbols. Sir Thomas Heath writing of the arithmetic of Nicomachus said 'If the verbiage is eliminated, the mathematical content can be stated in quite a small compass,' but Heath used modern notation and Arabic numerals. In the Wasps of Aristophanes one of the characters tells his father to do an easy sum 'not with pebbles but with fingers,' and Herodotus says that, in reckoning with pebbles, Greeks move left to right, Egyptians right to left, which implies vertical columns facing the reckoner.

The Greeks also developed a theory of geometry which remained more important than any other for nearly 2,000 years, and was the first deliberate development of a logical system in mathematics. In the third century A.D. an unknown writer jokingly used words of Homer intended for something else to describe mathematics:

Small at her birth, but rising every hour.
She stalks on earth and shakes the world around.

For, says Anatolius, Bishop of Laodacia, who quoted it, mathematics begins with a point and a line and forthwith it takes in the heaven itself and all things within its compass. If this was the Greek viewpoint at such a late date, is it possible that their geometry was not truly abstract and that the symbols of point and line were still partly coalesced with the abstract point and line?

The position of geometry and more generally spatial concepts in mathematics is not completely clear to me. In recent times all types of geometry have been given an analytical basis and freed from the logical difficulties such as those which used to worry schoolmasters teaching about congruent triangles by the method of superposition. I therefore ask myself whether geometry and spatial concepts are really part of the basis of mathematics or a field of application similar to mechanics, both terrestrial and celestial, or to games of chance. The reason for the traditional special position of geometry may be that in geometry the symbols are the objects themselves; the abstract point, line, and triangle are represented by a point, line, and triangle; what is more, so long as the geometry is plane geometry they can be drawn on a flat surface by pen or pencil on paper or in sand on the ground. When Greek geometry was being developed there was no good notation for dealing with numbers, and even in the 15th Century the solution of a cubic equation was described in geometrical terms and illustrated by a figure for lack of a good algebraic notation. In mechanics a comparable real life representation of motion could not be used to explain the theory; written symbols or geometrical figures were needed for communication. But if we ask whether the contributions of spatial concepts to modern mathematics are greater than those of other real life problems it is difficult to answer. Spatial thinking has led to the highly abstract theory of irrational numbers of Cantor and Dedekind, and permeates mathematical thought in almost all fields; the physical sciences have given rise to the calculus (not without the help of geometry), and statistics and probability have their basis in multitudinous practical problems.

Pfeiffer explains the situation well in relation to probability. Some of the salient points in his account are as follows: The history of probability theory (as is true of most theories) is marked both by brilliant intuition and discovery and by confusion and controversy. Until certain patterns had emerged to form the basis of a clear-cut theoretical model, investigators could not formulate problems with precision, and reason about them with mathematical assurance.

From what some people say it sounds to me as if quantum theory had not yet reached this stage, but it is certainly beyond my competence to form a valid judgment.

Pfeiffer continues by saying that although long experience was needed to produce a satisfactory theory, we need not retrace and relive the fumblings which delayed the discovery of an appropriate mathematical model. That is, a mathematical system whose concepts and relationships correspond to the appropriate concepts and relationships of the real world. Once the model has been discovered, studied, and refined, it becomes possible for an ordinary mind to grasp, in a reasonably short time, a pattern which took decades of effort and the insight of genius to develop in the first place. I note that Pfeiffer asserts that the most successful model of probability theory known at present is characterized by considerable mathematical abstractness.

J. Willard Gibbs wrote 'One of the principal objects of theoretical research in any department of knowledge is to find the point of view from which the subject appears in its greatest simplicity' and Bushaw says that one of the distinctive characteristics of modern mathematics is its way of taking old mathematical ideas apart like watches, studying the parts separately, and putting these parts together again in new and interesting combinations and studying these complications in turn. I believe that this process has contributed enormously to this simplification in mathematics itself, and so made it more readily available for applications. Mandelbrojt referring to the quotation from Willard Gibbs says 'Integration in function spaces provided such a point of view over and over again in widely scattered areas of knowledge and it gave us not only a new way of looking at problems but actually a new way of thinking about them.' Now one might call Fréchet the father of abstract spaces, and in the front of his book he puts a quotation from Hadamard's survey of functional analysis given in 1911. 'The functional continuum does not present any simple concept to our imagination. Geometrical intuition tells us nothing *a priori* about it. We are forced to remedy this ignorance and we can do it only analytically, by creating a chapter of the theory of sets for handling the functional continuum.' Elsewhere Hadamard wrote that the calculus of variations was nothing but the first chapter of functional analysis, and of his own work on the calculus of variations, hyperbolic partial differential equations, and certain other topics he said that he owed the greater part to his contacts with the physicist Duhem, through Duhem's book on hydrodynamics, elasticity, and acoustics and many conversations when they were both at Bordeaux. So we have a record here of the complete cycle from a physical basis through the calculus of variations to functional analysis and abstract spaces, and thence to a multitude of applications through the process of analyzing geometrical ideas and putting them together again in a most abstract new way to create function spaces.

A further variation on this pattern has become evident of recent years and that is the use of an auxiliary model consisting of various graphical, mechanical, and other aids to visualizing, remembering, and even discovering things about the mathematical model. The visual images of Hadamard, Hardy's cricket averages, and Littlewood's trajectories might be considered as auxiliary models, but of more universal significance are the analogue machines with electronic devices which simulate what happens in, for instance, fluid mechanics, or rather what corresponds in the mathematical model. We now have

(A) The real world of actual phenomena, known to us by various ways of experiencing these phenomena.

(B) The abstract world of the mathematical model which uses symbols to state relationships and facts with great precision and economy.

(C) The auxiliary model.

The transition from A to B is the formulation of real world phenomena in mathematical terms; the transition B to A is the interpretation of the deduction by pure mathematics from that formulation. Both these I consider to be thinking mathematically, but only the deductions inside B are mathematics. We may also think mathematically by moving from B to C which is a secondary interpretation, and then either back to B to confirm what C has suggested or from C direct to A.

As Pfeiffer points out, the value of both the mathematical model and the auxiliary model depends on how successfully the appropriate features of the model may be related to the 'real-life' situation. The models cannot be used to prove anything about the real world, although a study of it may help us to discover important facts about the real world. A model is not true or false; it fits or it does not fit. It is unsatisfactory if either (1) the solutions of the model problems have unrealistic interpretations, for instance, arbitrarily large quantities or arbitrarily fine differences, or (2) it is incomplete or inconsistent so that the mathematics produces contradictions. Many models fit amazingly well. Karl Pearson wrote 'The mathematician, carried along on his flood of symbols, dealing apparently with purely formal truths, may still reach results of endless importance for our description of the physical universe.

Until perhaps 100 years ago many scientists and mathematicians knew a bit of everything, and the mathematical formulation, as I said of Newton in particular, was done by someone who was a good enough mathematician to develop the mathematics to a considerable extent. This is particularly true of Sir Isaac Newton, but in these days of specialization the scientist or economist, or worker in close contact with the real world situation must do stage A → B. Sir Cyril Hinshelwood, former President of the Royal Society, said 'Scientists need to be taught mathematics as a language they can actually speak. It is of great importance for the scientist to be able to learn the art of formulating problems in mathematical terms which of course is a quite difficult job. You have to think very accurately and carefully about a problem before you can do it. You have to have practice in speaking the language of mathematics. It does not matter being an expert in differential equations. You can go to the expert for help in solving an equation. But you cannot expect the mathematician to do the translation into mathematics. There should be an early and rather intensive cultivation of the power of thinking about real things and the application of mathematical symbolism to physical ideas.' He went on to draw a parallel between learning simple French as a child and learning to express physical ideas in mathematics when the level of physics and mathematics reached are both elementary, so that the child becomes accustomed to the process by easy stages. Although he advocates, as I do, that the scientist should do the mathematical formulation, his words seem to imply an incomplete abstraction. In his mind the mathematical symbols were still representing their physical counterparts, not that this matters for a scientist who has access to an expert mathematician, but it is clear from Mandelbrojt's remarks on function spaces and Hadamard's remarks about the functional continuum that without complete abstraction on the part of some mathematicians we should lack some of the most expressive parts of the mathematical language used by scientists.

'Euclid's geometry was supposed to deal with real objects, whether in the physical world or in some ideal world. The definitions which preface several books in the *Elements* are supposed to communicate what object the author is talking about even though, like the famous definition of the point and the line, they may not be required in the sequel. The fundamental importance of the advent of non-Euclidean geometry is that by contradicting the axiom of parallels it denied the uniqueness of geometrical concepts and hence, their reality. By the end of the nineteenth century, the interpretation of the basic concepts of geometry had become irrelevant. This was the more important since geometry had been regarded for a long time as the ultimate foundation of all mathematics. However, it is likely that the independent development of the foundations of the number system which was sparked by the intricacies of analysis would have deprived geometry of its predominant position anyhow.'

Although it confirms my views on Euclidean geometry, it does not seem to recognize the geometrical origin of the theory of irrational numbers.

I also noticed that A. Aaboe in *Episodes from the Early History of Mathematics,* writes 'Even the oft repeated statement that the Egyptians knew the 3, 4, 5 right angle has no basis in available texts, but was invented about 80 years ago.'

References

1. S. Bochner, *The Role of Mathematics in the Rise of Science*, Princeton, 1966.
2. D. Bushaw, *Elements of General Topology*, Wiley, New York, 1963.
3. B. Datta and A. N. Singh, *A History of Hindu Mathematics*, Lahore, 1935.
4. H. and H. A. Frankfort, *The Intellectual Adventures of Ancient Man*, Chicago, 1946.
5. T. L. Heath, *A History of Greek Mathematics*, Vol. 2, Oxford, 1921.
6. S. Mandelbrojt, Les Tauberiens Généraux de Norbert Wiener, *Bull. Amer. Math. Soc.*, 72 (1966) 48–51.
7. O. Neugebauer, The Exact Sciences in Antiquity, *Acta Hist. Sci. Nat. Medicinalium*, Copenhagen, 9 (1951).
8. P. E. Pfeiffer, *Concepts of Probability Theory*, McGraw-Hill, New York, 1965.
9. J. Piaget, B. Inhelder, and A. Sjeminska, *A Child's Conception of Geometry*, Trans. by E. A. Lunzer, Basic Books, New York, 1960.
10. N. Tinbergen, *The Herring Gull's World*, Basic Books, New York, 1961.

Oh God! I could be bound in a nutshell and count myself king of infinite space, were it not that I have bad dreams. —W. Shakespeare (1564–1616)

Mathematical Invention

Jules Henri Poincaré

Henri Poincaré (1854–1912) deserves a place alongside the most original and creative mathematicians of all time. His work in so many fields has left a profound and indelible mark on many disciplines.

He was born in Nancy into a well-respected family of the province of Lorraine. His father Léon was a professor of medicine at the university of Nancy and his mother Eugénie Lannois came from the town of Arrancy. She is described as a pious, active and very intelligent woman.

Eugénie was devoted to her children, Henri and his sister, and under her tutelage Henri developed rapidly, learning to read at a very early age. As a youngster, he contracted diphtheria which left him weak for the rest of his comparatively short life. This weakness included poor eyesight and, being unable to discern the blackboard, it is said that he could mentally reconstruct the entire lecture from what he heard. An interesting aside is that he was ambidextrous. He had a cheerful personality and exuded friendliness. Endowed with a phenomenal memory, not surprisingly, he excelled at school. His originality manifested itself frequently; in particular at the age of 15 he wrote a play in verse about Jeanne d'Arc. During the Franco-Prussian War he taught himself German in order to keep abreast of the news. (Opponents of standardized tests will be interested to know that after achieving fame, he agreed to take the Binet test whose use was becoming more widespread; he performed abominably!)

His academic career was meteoric; after graduating from the École Polytechnique, he took the unusual step of entering the École des Mines, having a keen interest in mining engineering and geology. He worked in this area while completing his doctoral dissertation. His thesis was a significant piece of work in the theory of differential equations, and on its basis he was then appointed to a professorship at the Sorbonne where he remained until his untimely death in 1912 at the age of 58.

In 1881 he married Miss Poulin d'Andecy; she gave birth to five children. Poincaré was a devoted and attentive father and it is reported that he tutored them and rewarded them with prizes consisting of artistic drawings he had made earlier!

His originality included contributions to a breathtaking variety of disciplines ranging from number theory to topology, differential equations, probability theory, a wide miscellany of applied mathematics, and astronomy. Some writers credit him with having been a precursor of chaos theory, as well as the special theory of relativity.

In addition to his mathematical work, he turned his superb literary gifts to an elucidation of some issues in the philosophy of science, as well as to the question of creativity in these disciplines. His literary work became and remains very popular and on the basis of this he was appointed in 1908 to the Académie Française, the highest honor France can bestow upon a man of letters. In addition he received many other honors during his life.

His output was stunning and a measure of his influence is the fact that numerous items—papers and books commenting on his life and work—continue to be published. In addition, his name is associated with many mathematical concepts and discoveries. His legacy is, and will undoubtedly remain, one of the richest in the history of science.

Editor's Preface

Poincaré had a lifelong interest in the humanities and the philosophy of science and wrote extensively on these matters. Here, this interest finds its expression in his meditations on what he calls *mathematical invention* (many would call it mathematical discovery or creation).

He begins by pointing out what he sees as a curious fact: Why is it that so many people of "sound" mind have difficulty understanding mathematics? It is, after all, a simple concatenation of syllogisms. That most people cannot invent mathematics is not astonishing, but that so many are incapable of understanding mathematics is a mystery, says Poincaré. It is a mystery that confronts all who teach mathematics. Reams have been written on the subject of mathematical pedagogy but these have had, for the most part, limited success.

We believe that Poincaré overstates the ability of people to understand logical inferences—one has only to witness abundant examples in the real world of oral or written expression!

Leaving this matter aside, he now turns his attention to mathematical invention and cites examples from his own experiences. These generally take the following form: He works diligently on a problem without success. Some time later he is struck by a sort of epiphany, more or less out of the blue, and the solution comes to him—he has only to verify the details. This same experience, he says, is not at all uncommon among mathematicians. What has happened between the time he has consciously worked on the problem and the sudden revelation? Poincaré postulates that the subconscious is at work in the background, so to speak, forming innumerable combinations out of the conscious facts. It would take a lifetime to examine the multitudinous combinations so formed. What leads then, to the combination which proves to be fruitful? To answer this question, Poincaré makes the interesting assertion that the successful mathematician is endowed by nature with a strong sense of the harmony and beauty of mathematical laws. It is this sense that enables the mathematician to select expeditiously the most felicitous from among the myriad potential combinations that could be formed by what he calls the subliminal mind. In the absence of this quality the process of selection could be interminable

This endowment, that is, the esthetic sense, is essential and should be accompanied by a good memory and good powers of concentration. Being possessed of the second and third are sufficient to learn mathematics but the first quality is necessary for its creation. How it can be recognized or assessed is as elusive as, for example, identifying creativity in literature and the arts. Poincaré possessed it to a high degree and this found expression in his many beautiful and original ideas.

An aside is called for. Although the word "subconscious" was used in the 19th c. it is interesting that Poincaré uses the words "subconscious" and "subliminal" in a sense similar to that used by Freud. The *Oxford English Dictionary* dates Freud's use to 1911 while the date of this article is 1908.

Note. The editor has elected to translate the French word "invention" as the English word "invention." There is ample historical justification for the use of the word in this context, where many authors use the word "creation." "Invention" has the approbation of the O.E.D.

The genesis of mathematical invention is a problem which should inspire the most lively interest among psychologists. It is a process in which the human mind appears to borrow least from the external world, where it does not act, or does not appear to act, except by itself or upon itself, so that in studying the process of mathematical thought, what we hope to attain is that which is the most essential element in the human mind.

The need for this study has long been understood, and several months ago, the journal called *Enseignment Mathématique*, which is edited by Messrs. Laisant and Fehr, undertook a study on the mental processes and the working habits of various mathematicians. I had already composed the main points of my lecture when the results of this study were published; I was not able to use the results at all. I shall limit myself to saying that the majority of those sampled confirm my conclusions: I do not say all of them, for when one examines an entire class, one cannot flatter oneself into believing that the outcome will be universal.

A first fact should astonish us, or should have astonished us, if we were not so used to it: How does it happen that there are individuals who do not understand mathematics? If mathematics invokes nothing but the rules of logic, those which are accepted by all normal individuals and if their evidence is based on principles which are common to all humans and which no one can deny without seeming foolish, how does it happen that there are so many people who are totally resistant to mathematics?

That not everyone is capable of invention is not in the least mysterious. That not everyone can retain a proof that they once learned is also understandable. But that not everyone can understand a mathematical argument when we explain it, that is what is most surprising if we think about it. Besides, those who cannot follow this reasoning except with great difficulty are in the majority: that is undeniable and the experience of secondary school teachers would certainly not contradict me.

And there is more: how is error possible in mathematics? A sound intelligence should not make an error in logic, and yet there are responsive minds who do not stumble in a short argument, such as those with which we deal in the ordinary course of our lives, but who are incapable of following, or repeating without error, proofs in mathematics that are longer, but that are, after all, nothing more than an accumulation of small steps entirely analogous to those that they perform easily. Is it necessary to add that good mathematicians themselves are not infallible?

The answer seems to suggest itself. Imagine a long series of syllogisms such that the conclusions of the first serve as premises for the following ones; we should be capable of understanding each of these syllogisms, and it is not in the transition from the premises to the conclusion that we risk error. But from the time we encounter a proposition for the first time, as the consequence of a syllogism, till the time that we encounter it as the premise of another syllogism, occasionally a long time has elapsed and we should have unraveled numerous links of the chain; it could then have happened that we could have forgotten it, or what is more serious, that we have forgotten the meaning. It could happen that we replace it by a proposition which is a bit different, or that keeping the same assertion, we attribute a slightly different meaning to it, and it is thus that we are subject to error.

Often the mathematician needs to use a statement; naturally he starts by proving the statement. At the time this proof was fresh in his mind, he understood perfectly the sense and significance, and he did not risk altering it. Moreover he memorized it and did not apply it except in a routine manner. Then if his memory had misled him, he could have applied it incorrectly. It is thus, to take a simple and mundane example, that we sometimes make errors in calculation because we have forgotten our multiplication table.

On this point, a special mathematical aptitude is based on a reliable memory, or is the result of a prodigious ability of concentration. It is a characteristic analogous to a bridge player who remembers all the cards played; or to go up a step, to a chess player who can envisage a large number of combinations and keep them in his memory. A good mathematician should, at the same time, be a good chess player and conversely, he should be good at numerical calculations. Certainly this sometimes happens. Thus Gauss was a mathematical genius and a precocious and accurate calculator.

But there are exceptions, or rather I am mistaken: I cannot call these exceptions for the exceptions are more numerous that those that conform to the rule. It is Gauss on the contrary, who is the exception. As for me, I am obliged to confess that I am absolutely incapable of doing addition without mistakes. I should also be a very poor chess player. I should reason that in playing in a certain way, I expose myself to a certain danger. I review many other moves that I reject for other reasons, and I end up by playing the move which I first examined, having forgotten, in the meantime, the danger which I had foreseen.

In a word my memory is not bad, but it would be inadequate to make of me a good chess player. Why then does it not fail me in a mathematical reasoning where most chess players would be lost? It is evidently because it is guided by the general progression of the argument. A mathematical argument is not simply a juxtaposition of syllogisms: it is syllogisms placed in a certain order, and the order in which these elements are placed is more important than the elements themselves. If I have the sense, the intuition so to speak, of this order, in such a way that I perceive at a glance the totality of the argument, I no longer fear that I shall forget one of the elements. Each one will be placed in the frame which was prepared for it, and without my need to make any effort at remembering.

It thus seems to me, that in repeating an argument that I have learned, I could readily have discovered it myself: or rather, even if this is an illusion, if I were not clever enough to have created it myself, I rediscover it myself, to the extent that I repeat it.

We perceive that this sense, this intuition of mathematical order, which enables us to perceive harmonies and hidden relations, cannot be present in everyone. Some would not have either this delicate sense, which is difficult to define, nor the power of memory and attention above the average, and thus they will be incapable of understanding somewhat advanced mathematics: these are in the majority. Others will have this sense to a small degree, but they will be endowed with an uncommon memory and with a large capacity for concentration. They will learn by heart details one after another: they will be able to learn mathematics and sometimes learn to apply mathematics but they will not be able to create mathematics. The others finally, will possess, to a greater or lesser degree, a special intuition of which I have spoken, and thus not only can they understand mathematics, even though their memories are not unusually good, but they will be able to become creators and seek to create with more or less success, according to the extent that this intuition is developed in them.

Indeed what is mathematical invention? It does not consist of making new combinations of objects which already exist. Anyone can do that: but the combinations that one can make would be finite, and the majority would be without any interest. Inventing consists precisely in not constructing combinations that are useless, but in constructing those which are useful and these are a small minority. Inventing means discerning, means choosing.

How this choice should be made I have explained elsewhere. The mathematical facts that are worthy of study are those that, by their analogy with other facts, are susceptible of leading us to a knowledge of a mathematical law, in the same way that experimental facts lead us to a physical law. It is those that reveal unsuspected relations among other facts that were long known but which were thought to be unrelated to one another.

Among the combinations that we shall choose, the most fruitful will be those that are borrowed from domains that are far removed. I do not mean that to invent, it suffices to invoke objects which are as disparate as possible: most of the combinations so formed will be entirely sterile: but several among them, very rare, are the most fruitful of all.

Invention I have said is choosing, but the word is not exactly justified. It brings to mind a buyer to whom we offer a large number of samples; he examines them one after the other in order to make his choice. In our case the number of choices is so large that an entire lifetime would not suffice to examine them all. This is not the way that these things take place. The sterile combinations are not even considered by the inventor. In the realm of his conscience there will never appear any but combinations that are useful, and several others that he discards, but that have some of the characteristics of the useful ones. Everything takes place as though the inventor were an examiner of the second degree, who need no longer examine any but the combinations that are admissible after having been subjected to a first examination.

But that which I have asserted up to now is what we can observe or infer upon reading the works of mathematicians, with the condition that we read with a certain deliberation

It is time to go further and to see what takes place in the mind of a mathematician. For that, I believe that the better thing to do is to recall some personal observations. However I shall limit it in only describing how I wrote my first memoir on fuchsian functions. I ask your pardon since I shall use several technical expressions; but these should not frighten you since you have no need to understand them. I shall say, for example, that I found the proof of such and such theorem in such and such circumstances. This theorem will have a monstrous name that many among you will not know, but that is of no importance. What is interesting to the psychologist is not the theorem but the circumstances.

For fifteen days, I tried to prove that there could not exist any functions analogous to those I have since called fuchsian. I was then very ignorant; every day I sat at my work table where I passed an hour or two trying a large number of combinations but I could not come to any result. One evening, I drank some black coffee contrary to my custom. I could not sleep and ideas surged in large numbers. I sensed them running into one another to the point where two among them locked, so to speak, to form a stable combination. In the morning, I had established the existence of a class of fuchsian functions, those derived from the hypergeometric series. I had only to verify the results, which took me only several hours.

I then wanted to represent these functions as the quotient of two series. This idea was a completely conscious and thoughtful one; the analogy with elliptic functions guided me. I asked myself what ought to be the properties of these series if they exist, and I arrived without difficulty at forming the series which I called thetafuchsian.

At this time I left Caen, where I lived, to take part in an outing sponsored by the School of Mines. The meanderings of the trip made me forget my mathematical work. When I arrived at Coutances we embarked on a bus for some trip or other. At the moment that I put my foot on the step, the idea came to me, without anything of my previous thoughts appearing to have prepared me, that the transformations that I had used to define fuchsian functions were the same as those of non-euclidean geometry. I did not verify it. I would not have had the time since as soon as I sat down in the bus, I resumed a conversation I had started, but I had immediately a complete certainty. Returning to Caen, with a rested mind I verified the result.

I set out to study arithmetic questions without significant results appearing and without suspecting that that would have the least relation with my previous results. Disgusted with my lack of success, I decided to pass several days at the seaside, where I thought of other things. One day, walking along a cliff, the idea came to me, with the same characteristic of brevity, of an immediate suddenness and certainty, that the arithmetic transformations of ternary indefinite quadratic forms were identical with those of non-euclidean geometry.

Upon returning to Caen, I thought about this result and I derived consequences: the example of quadratic forms showed me that there are fuchsian groups other than those corresponding to hypergeometric series. I saw that I could apply to them the theory of thetafuchsian functions other than those derived from hypergeometric series, the only ones I knew up to the present. I naturally decided to form all of these functions: I formed a systematic assault and I eliminated one after the other: there was one however that resisted my efforts and whose elimination entailed that of the entire corpus. But all my efforts served, at first, to enable me to understand the difficulty. That at least was something. All this work was entirely conscious.

After that, I departed for Mount Valerian, where I was to do my military service. I therefore had other preoccupations. One day in crossing a boulevard, the solution of the difficulty that had hindered me, appeared to me all at once. I did not seek at once to go more deeply into it and it was only after my service that I took the question up again. I had all the pieces, I had only to put them together and to order them. I then composed my definite memoir without difficulty.

I shall limit myself to this one example. It is useless to elaborate. As far as my other researches are concerned, I should only give analogous accounts: and the observations reported by other mathematicians in the investigation in *L'Enseignment Mathèmatique* can only confirm these.

What will first strike you are the sudden appearances of clarification that are obvious signs of long unconscious prior work; the role of this unconscious work, in mathematical research, appears undeniable to me and we should find traces in other cases where it is less evident. Frequently when we work on a difficult question, we have no success the first time we undertake the task: then we take a long or short rest, and we sit at our desk anew. During the first half hour we continue to find nothing. Then all of a sudden, the decisive idea comes to one's mind. One can say that the conscious work had been more fruitful because it had been interrupted and that the rest had given the mind power and freshness. But it is more likely that the rest was filled with unconscious work and that the result of this work was revealed to the mathematician, more or less in the manner I have cited, except that the revelation, instead of coming to light during a walk or a journey, manifests itself during a conscious period of work, but independently of this conscious work, which plays the role of a triggering action. It acts as though it were a nudging stimulus that had excited the results already acquired during the period of rest. These had remained in the subconscious, but now revert to the conscious form.

There is another remark to be made on the subject of the conditions of this subconscious work. That is, that it is not possible, and in any case, that it is not fruitful, unless it is, on the one hand preceded, and on the other hand followed, by a period of conscious work. Never (and the examples that I have cited are sufficient proof) are these sudden inspirations produced except after several days of voluntary effort, which have appeared fruitless and which one believed were of no value, where it appeared that we have taken a totally false path. These efforts were, therefore, not as sterile as we had thought: they put our subconscious into motion, and without these, it would not have been set in motion and would not have produced anything.

The necessity of the second period of conscious work, after the inspiration, can better be understood. It is necessary to put to work the result of this inspiration, to deduce immediate consequences, order them, and write the proofs. But above all it is essential to verify them. I spoke of the sense of absolute certainty which accompanies this inspiration; in the cases cited, this sense was not a deceiver, and for the most part, that is the case: but one should be careful not to assume that this is the case without exception. Often this sense deceives us, although it is not less vivid, and we perceive it only when we seek to put the proof on a sound footing. I observed this fact, moreover, for the ideas that came in the morning or the evening while in my bed, in a semi-hypnotic state.

Such are the facts, and here are the thoughts that are laid upon us. The word *subconscious*, or as we say the *subliminal me* plays an important role in the discovery of mathematics. That follows from all that has been said. Ordinarily, we consider the *subliminal me* as purely automatic. However we have seen that mathematical work is not a purely mechanical work that we can relegate to a machine, however perfect we may assume it to be. It is not a matter only of applying rules, to put together as many combinations as possible according to certain fixed laws. The number of such combinations is extremely large, useless and cumbersome. The true work of the researcher is to choose combinations from among these, in such a way as to eliminate those which are useless or rather are not worth forming. And the rules that should guide this choice are extremely fine and delicate: it is almost impossible to express them in a precise language; rather *they* sense that they cannot form themselves. How, under these conditions, can we imagine a sieve that is capable of being applied mechanically?

And now a first hypothesis suggests itself to us: the *subliminal me* is not inferior to my conscious self. It is not purely automatic; it is capable of discernment. It has a tactile sense; it has sensitivity. It knows how to choose; it knows how to guess. It knows how to guess better than the conscious me since it has succeeded where the conscious has failed. In a word, is the *subliminal me* not superior to the conscious me? You can readily understand the importance of this question. Mr. Boutroux, in a lecture he gave here a couple of months ago, showed you how it intervened in different circumstances and what consequences an affirmative answer entailed.

Is this affirmative answer imposed by the facts which I have given? I affirm that, on my part, I do not accept it without repugnance. Let us look again at the facts and see if they do not require another explanation.

It is certain that the combinations that come to mind, in a sort of sudden illumination, after a somewhat prolonged unconscious effort, are generally useful and fruitful combinations that resemble the result of a first attempt. Does it follow that the *subliminal me*, having guessed, by a delicate intuition, that the combinations might be useful, formed only these, or it formed many others that were devoid of interest and that remained in the subconscious.

In this second way of viewing it, all the combinations will be formed as a consequence of the automatism of the *subliminal me*, but only those that are interesting penetrate into the field of the conscious. And that is also very mysterious. What is the cause that results in the fact that among the thousand products of our subconscious activity, there are some that are provoked to cross the threshold, while others remain on the side? Is it a simple act of chance that endows them with this privilege? Obviously not: among all the excitations of our senses, for example, only the most intense hold our attention, so long as this attention has not been diverted by other causes. More generally, the privileged subconscious phenomena, those capable of becoming conscious, are those that directly or indirectly affect the most profound of our sensibilities.

One can be astonished at the idea of invoking sense impressions in connection with mathematical proofs, that on the face of it can only be of interest to the intellect. This would be neglecting the perception of the beauty of mathematics, the harmony of numbers and forms and the elegance of geometry. It is a true esthetic sense that all true mathematicians recognize. And it is indeed there that we find sensitivity.

What are the mathematical objects however, to which we attribute this quality of beauty and elegance and which are capable of developing in us a sort of esthetic emotion? They are those of which the elements are harmoniously distributed, is such a way that the mind can, without effort, grasp the entire set while probing the details. This harmony is both a satisfaction for our esthetic needs and an aid for the mind that then verifies and guides. And at the same time, in putting before us a well-ordered object, it enables us to sense a mathematical law. However we have said above that the only mathematical facts worthy of our attention and capable of being useful are those that have the potential for discerning a mathematical law. Consequently, we reach the following conclusion: the useful combinations are precisely the beautiful ones, I mean to say those that are capable of invoking this special sensibility which all mathematicians know, but which lay persons ignore to the point where they are tempted to smile.

What have we concluded? Among the very large number of combinations that the *subliminal me* has blindly formed, almost all are without interest and are useless, but because of this fact they have no effect on the esthetic sense. The conscious mind will never recognize them. Only a few are harmonious and consequently both useful and beautiful. They are capable of affecting this special sensibility of the mathematician of which I have spoken, and the mathematician, once excited, calls our attention to them and gives them the opportunity of emerging to our conscious mind.

That is only a hypothesis, and here is an observation which could confirm it: when a sudden illumination envelops the mind of a mathematician, it often happens that it does not deceive him. But it is also the case that sometimes it does not support the test of verification; well then we observe that almost always this false idea, had it been correct, would have pleased our natural instinct of mathematical elegance.

Hence it is this special esthetic sense that plays the role of a delicate sieve of which I spoke above, and that enables us to understand why the person, who is deprived of this sense, will never become a true discoverer of mathematics.

However not all the difficulties have disappeared: the conscious me is narrowly prescribed, while we do not know the limits of the *subliminal me*. And that is why we are not too reluctant to suppose that it was able to form more different combinations in a short span of time than the conscious mind could form in its entire lifetime. Limits exist however; is it reasonable that it could form all possible combinations, of which the number beggars the imagination? That is nevertheless necessary, since if it produces only a small part of these combinations, and if it does it at random, there would be little chance that the good one, that which we should choose, would be found among them.

Perhaps we should seek the explanation in that period of work that always precedes a fruitful unconscious period of work. Permit me a gross comparison. Let us represent the future elements of our combinations as something similar to hooked atoms. During the complete period of rest of the mind, these atoms are immobile. They are, so to speak, hooked to the wall. This complete rest can continue indefinitely without these atoms colliding, and consequently without a single combination being produced by them.

By contrast, during a period of apparent rest and unconscious work, several of them are detached from the wall and set into motion. They streak throughout space—I was going to say in the region where they are enclosed, as for example a cloud of gnats, or if you prefer a more learned comparison, as gas molecules in the kinetic theory of gases. Their mutual collisions could thus produce new combinations.

What will be the role of the preliminary conscious work? It is evidently to mobilize some of these atoms, to unhook them from the wall and to put them in motion. We first believe that we have done nothing worthwhile because we have moved the elements in a thousand different ways in order to reassemble them and we have not found a satisfying assemblage. But after this agitation which was imposed by our will, these atoms will not return to their primitive state. They continue their dance at will. However our will has not chosen them at random, they follow a perfectly determined objective; the mobilized atoms are not any atoms; they are those that have a reasonable chance of attaining the solution that was sought. Those atoms that are mobilized will be subject to collisions, which will force them to enter into combinations, either among themselves or among atoms that were stationary and that come into collision in their trajectory. I ask your pardon again: my comparison is crude, but I do not know how I could otherwise make my thoughts understood.

However that may be, the only combinations that have a chance of being formed are those where at least one of the elements is one of the those freely chosen by our wishes. However it is evidently among these that can be found that which I have called, a moment ago, a good combination. Perhaps there is in this a means of attenuating what was paradoxical in the primitive hypothesis.

Another observation. It never happens that the subconscious work provides us in complete form the result of a somewhat long calculation where one need only apply fixed rules. One could think that the subconscious me, entirely automatically, is particularly adept at this type of work, which is, in a way, exclusively mechanical. It is as though in thinking in the evening about two numbers, we can hope to find the product on awakening, or else an algebraic calculation, or a verification, for example, could be made subconsciously. That is not the case as observation shows. All that we can hope from these inspirations, which are the results of unconscious work, is that they are the starting points for similar calculations. As for the calculations themselves, it is necessary to carry them out during the second period of conscious work, that which follows the inspiration, that period where we verify the results of the inspiration and where we draw consequences. The rules of these calculations are strict and complicated: they require discipline, attention, the mind and consequently the conscious. In the subconscious me, there reigns what I have called freedom, if we can give this characterization to the simple absence of discipline or to confusion born of chance.

I shall make a final remark: when I recounted above several personal observations, I spoke of a night of excitement, when I worked despite myself. The cases when this happens are frequent, and it is not necessary that abnormal cerebral activity be caused by a physical excitation as in the case I cited. Well then it seems that in this case, we help ourselves to our subconscious work, which has become partially perceptible to the excited conscious, and which has not, accordingly changed its nature. We vaguely take into account that which distinguishes the two mechanisms or, if you like, the method of work of the two "mes," i.e., the conscious and subliminal. And it seems to me that the psychological observations that I have been able to make, appear to confirm, in their general characteristics, the point of view which I have proposed.

Certainly they are essential, because despite everything, they are and remain hypothetical. Interest in the matter is so very great and I do not regret having propounded them to you.

Mathematics presents the most brilliant example of how pure reason may successfully enlarge its domain without the aid of experience.

—Immanuel Kant (1724–1804)

Thoughts on the Heuristic Method

Jacques Salomon Hadamard

Jacques Salomon Hadamard (1865–1963) was born in Versailles. He lived a long and fruitful life marked by intellectual triumph and some personal adversity. The family originated in Metz and it is thought that they were among the Jews who came with the Roman legions to Lorraine. He was a precocious child who excelled in all subjects except mathematics. At this level "mathematics" means "arithmetic" and it is likely that he found the routines a bit boring! He eventually graduated from the Lycée Louis-le-Grand—a noted secondary school attended by many celebrated French scholars. He then entered the École Normale Supérieure. Here he studied, among others, with Hermite, Darboux and Picard, and was awarded a doctorate in 1892. In the same year he was awarded the Grand Prix des Sciences Mathématiques for his work on entire functions.

He had already demonstrated his remarkable creativity and to this was added a jewel in his crown—a proof of the prime number theorem. This deep result, foreseen by Gauss, was profoundly explored by Riemann who led the way, and the conjecture was finally resolved by Hadamard and Charles de la Vallée-Poussin more or less simultaneously. During the rest of his long mathematical career he produced one outstanding result after another and his name is attached to a variety of important mathematical concepts and results. His influence on his contemporaries and successors was profound. One of his more popular creations was a book entitled *The Psychology of Invention in the Mathematical Field.* In it he explores the creative process in mathematics. It had a wide readership and was translated into many languages.

His personal life was not the tranquil one sometimes associated with a scholar. In 1892 he married Louise-Anna Trenel, the daughter of one of his mother's friends. She was a gifted musician and "a generous and intelligent woman who supported and helped him throughout his career." They had two daughters and three sons. Tragedy struck when two of the sons were killed in World War I; this tragedy was compounded by the death of the third in World War II.

Hadamard became involved in the Dreyfus affair. This sordid episode involved allegations of espionage by Alfred Dreyfus and undercurrents of anti-Semitism. Hadamard was active in a successful movement to exonerate him. Hadamard's interest in the case was undoubtedly enhanced by the fact that Dreyfus' wife was a distant cousin of Hadamard's.

After the First World War, Hadamard became an energetic campaigner for peace especially with the Committee for Human Rights. He wrote articles and pamphlets that were widely read.

With the rise of Hitler in Germany, Hadamard was involved in freedom movements and when France fell in 1940 he escaped to the USA. He stayed there until France was liberated. His next trip to the USA was in 1950 to the International Congress of Mathematicians where he was made honorary president of the Congress. His admission to the USA was not without difficulty because of his past and then current political activities together with the suspicious mood at the State Department. He died in Paris in 1963.

Editor's Preface

This article was published in 1905 and was in response to a conference held the preceding year on questions of pedagogy. It discusses the *heuristic* method of teaching that is derived as a method, as Hadamard tells us, from the "Socratic method." This latter method consists of a series of questions, designed to elicit from the student, knowledge that, according to Socrates, is implicitly known to any rational being. The discussion is particularly relevant in view of the debate concerning contemporary reform movements in mathematical instruction.

The decline in students' performance in mathematics courses at all levels, has given rise, in the past several decades, to a wide variety of proposals to remedy this problem. These range in applicability from preschool to college and the methods, superficially at any rate, vary widely. There is, in particular, a vigorous movement under the rubric "calculus reform." What this means or what it entails, we leave to the reader. Mathematical pedagogy has had a history of instructional innovation but few methods differ as dramatically as those in place today.

In the essay by Hadamard, he discusses what he calls the "heuristic method." This, as Hadamard tells us, is derived from the Socratic method to which Socrates gave the name "maieutics." Socrates believed that any rational person, by which we presume he meant any person with a minimal faculty of reasoning, is latently in possession of knowledge and it is only a matter of making the person aware of it. The etymology of the word maieutics is from the Greek word meaning "midwife." The teacher acts as a midwife to help in the birth of knowledge and ideas.

"Heuristics," according to Hadamard, is a modification of the method of "maieutics" and is designed to be applied to larger groups than those associated with Socrates. The word is derived from the Greek word meaning "discovery." While we are at it, let us note that the word "educate" is derived from the Latin "to lead out." Thus all three words have allied connotations, linguistically and pedagogically.

No one can doubt that the Socratic method is a very powerful pedagogical tool but alas, it is not realistic for mass education. It requires a person well trained in the technique, to work with an individual or small group and to be unencumbered by societal and disciplinary forces. Heuristics offers an alternative but a limited one.

Nevertheless, an examination of many recent innovations leads to the conclusion that maieutics or heuristics is present tacitly or explicitly in these methods and indeed, some research workers essentially advocate the use, in so far as practicable, within the context of the learning environment.

In this era, the most widely recognized advocate of heuristics is the renowned mathematician George Pólya. He wrote extensively on "problem solving" and it is under this name that the method has gained widespread popularity and usage. The nature of the educational environment, however, decrees the extent to which it can be put into practice. Knowledge acquired in this manner, however, makes a permanent contribution to the student's education, quite independently of the subject matter being taught.

Hadamard's thoughts on the subject are given below.

The discussions which took place in 1904 at the Pedagogic Museum, and on which Ascoli reported shortly before his premature death, raised a series of questions on the teaching of science. It seems to me that it is proper to return to the one posed by Mr. Marotte, viz., the heuristic method.

Let us recall that this method is derived in principle from that which Socrates used and to which Socrates gave the name "maieutic." "In principle," since the Socratic method is our ideal model because it is not, in reality, a true heuristic method. But detached from this particular form, the essential principle holds with all its importance, and the efforts made abroad recently to apply the method deserve our attention.

"Here," says Mr. Marotte, "the professor lectures almost constantly, while the student remains passive. In Germany, the professor is a guide and the student is active.

"The entire class is involved; very short segmented questions are posed, going quickly from one student to another in order to maintain their attention. These questions are conducted in such a way, either to have the student discover the mathematical property to be proved or to derive from an experiment conducted before them, the physical law to be perceived."

Such a manner of proceeding obviously applies to the physical or natural sciences as well as mathematics, and not only the sciences but all possible branches of instruction. Applied to the experimental sciences, the method often carries here the American name "rediscovery." But the heuristic or rediscovery method represents, on the whole, a single objective: permit the student to discover the largest possible number of truths that the teacher has in mind.

With that in mind, what are the advantages and what are the objections?

Instead of seeking the advantages that the method has, it is essential first to look for those advantages it does not possess, because taken literally, the method would tend simply to make discoverers of all the students. It is necessary therefore to show that we are not in any sense speaking to an Archimedes or a Newton—we are aspiring to something possible, not utopian.

Since the system functions in Germany, we have to believe that it is viable, and the ideal which it proposes can be achieved, at least to a certain point of view and to a certain degree, both of which remain to be defined.

To whatever point of view and to whatever extent, we thus stimulate the activity of the student. This advantage, the first among many, cannot fail to entail others, which are far from being negligible. That is what Mr. Marotte determined:

"The classes conducted by this method, are much more animated than those conducted by the method of exposition.

"The use of the heuristic method is one guarantee that the instruction will not surpass the capacity of the student, and that the content is assuredly understood and retained."

And the objections?

There are objections of all sorts.

Should we recall those that we have made to remarks of Mr. Drucker when he lectured on a form of the method for instruction in the natural sciences? Certain of these were filled with humor: to make the students work and to train them in making observations, "it is less tiring than giving lectures"; others were filled with common sense: if we modify the process, according to the number of students, "this would make two methods of instruction." These remarks and others of the same persuasiveness, and others less so, were stated seriously at the Pedagogic Museum. They were said without laughter and were heard without weeping.

Serious objections exist—we hasten to come to them. It is essential, in my opinion, to take them seriously into account, if we do not wish to arrive at a "checkmate" for which the method should not be held responsible. Concerning mathematics, these objections were presented in a particularly interesting, complete and convincing manner by Mr. Duran.

Because, let us say at once, I cannot agree with the conclusions of the author. Mr. Duran is not, if you will, entirely opposed to the use of the heuristic method. But it is, in a way, a great concession that he is inclined to allow it (I hope he will pardon me if, perhaps, I exaggerate his thought) as a distraction for the audience, as a means—neither better nor worse than others—by which to vary one's task. He claims that continuous usage is hazardous.

If by "continuous" he means constantly and exclusively, then I agree with him. Two of the drawbacks that he cites seem to me, indeed, defective in these circumstances: the considerable time they require, the confusion they generate and the harm they bring to clarity.

The two objections are in reality but one. If one must eliminate the loss of time engendered by the "rediscovery" method from an entire scientific course, it is mainly because the audience would risk losing sight of the progression of ideas, or even losing, at the end, that which was said at the beginning.

But if the use of the heuristic method be made by slow degrees, I do not believe that henceforth it should alone dominate and result in the total disappearance of didactic instruction (I designate by "didactic" all that which is normally described as instruction, as opposed to "heuristic") nor do I suggest that it should play a purely episodic role; on the contrary we should compel ourselves to use it continuously, i.e., to call it up constantly and on all occasions.

The judgement that we are led to adopt on this new method depends, in fact, on the aim we have in mind. If the final end of instruction is the knowledge of such and such part of the program, one needs only to search for the simplest and fastest method for acquiring this knowledge, and reject all other methods or at most use these methods from time to time "for a change." It has not been shown that the heuristic method answers to this ideal. But the heuristic method is not, to my knowledge, the only one we should aim for: it is not that method which we have in mind when we speak of the educational character of instruction. However, in my opinion, this educational character depends in large part on the heuristic method.

We seek, in a word, to use this method, not because it permits us to acquire the same knowledge more easily, but because once acquired, the knowledge teaches us to reason more effectively.

While, as we have said, not being more inclined than Mr. Duran to agree to an excessive usage, which would compromise the clarity of the lesson, we do not share his concern for certain lessons which are "perfectly clear, concise, well-planned" and which be believes are disappearing. If the heuristic method sometimes leads to a reduction of the number, I should also not make a serious complaint against it. Who among us does not recall such lectures and has not sensed the drawbacks? Congealed in their perfection, they belong, in summary, to the category of "fossilized ideas," of which Mr. Marotte speaks. The students whom we benefit are like travelers who are too well guided, who assume they know a country because they know that which we wanted them to see—without troublesome incidents, with certainty, and without loss of time.

If we think, and I am not against it, that they too have advantages and that they are appropriate to fascinate the mind, it is these methods that we offer to our students from time to time for a change.

Moreover is it so evident that mathematics is more easily and more definitely understood by these model lessons than by the heuristic method? That depends on what we mean by "understand."

"This word " said Mr. Poincaré at the Pedagogic Museum, "does it have the same meaning for everyone? Does understanding the proof of a theorem consist of examining each step of the syllogisms that comprise the proof and certify that they are correct, in accordance with the rules of the game? In the same way, understanding a definition, is it merely to recognize that we already know the meaning of all the terms used and certify that they imply no contradiction?

"Yes for certain individuals; when they have come to this conclusion, they will say: I have understood. No for the majority. Almost all are more exacting: they want to know not only if all the syllogisms of a proof are correct, but why they concatenate in one order rather than another. As long as it seems to them to be capricious and not by an intelligence constantly aware of the end to be reached, they do not believe they have understood."

I decided to quote at length these words. They can serve as an epilogue to the present article. It is clear that Mr. Poincaré has put his finger on the crux of the matter. Those whose state of mind he thus describes, and who have no doubts themselves, justifiably want to understand something other than what our lesson offers in general. These often represent the majority.

I hope that the heuristic method will bring a remedy to this ailment which afflicts mathematical instruction in the secondary schools. What I know is that all the other methods currently in use do not seem in the least bit capable of achieving this objective.

I hope I have shown that it will not suffice, if we want to abandon this method, to determine that its use gives rise to difficulties: we must show that these difficulties are insurmountable or, at the least, are sufficiently serious as to force us to give up the principal advantages which it brings.

Naturally, this is not a reason, if the method is considered important, to be involved in these difficulties and to solve them. I believe with Mr. Durant, that up to the present we have not looked ahead sufficiently. We have engaged in talking about the heuristic method; we have discussed its advantages and drawbacks without agreeing on our conception of its mechanism. I do not think however, that it can be improvised; and for my part, for the many years that I have used it, experience has led me unceasingly to modify, in one point or another, my method of proceeding. That is to say, I do not have the pretension of announcing definite rules; I should feel that I have achieved my goal if the thoughts I present will generate others.

To begin with, how will the students resolve questions that cannot be elucidated without the genius of an Archimedes, a Pythagoras or a Newton? We have seen that that is the objection that comes immediately to mind. Mr. Durand who poses the objection provides, implicitly, an answer when he compares Mathematics to a staircase with uneven steps, some of which are too high to be reached without the help of a teacher.

The conclusion resounds with confirmation: if the steps are too high, it is necessary to use other, smaller steps. We should segment the questions in such a way as to fit the intelligence of those to whom they are presented.

For those who feel compelled to operate prudently, or for whom the level of the class forces them to do so, there is no limit to this fragmentation. The more the listeners need to be managed, the more one needs to simplify the questions. None of these dissections would reach the point where they are not beneficial to the mind. For example the pure and simple management of an algebraic calculation whose progression has been indicated by the professor is already a profitable exercise. Several of these calculations can be done by analogy with those normally assigned for homework. I do not see any drawback to the fact that one of these assignments is part of the course; I see, on the contrary, all sorts of advantages to convey to the students that this part does not count as far as the difficulty is concerned since it does not differ from those they habitually do by themselves.

But, in general, we can soon make a further step, and, instead of proceeding with the calculations for which the application of the rules presents no difficulty whatsoever, we can propose others where this application requires a certain effort of concentration, where for example it hides snares such as those where the interjection of auxiliary unknowns often holds back beginners. The heuristic questions then most often limit themselves to yelling "trap."

Tailoring the steps in slopes which are to be climbed: that will be our first preoccupation, and that evidently, should be independent of the science that we have to teach.

Concerning mathematics, we can reassure ourselves especially simply, that the breaking up is already done for the most part in the present status of teaching. Not a single question is presented in those terms in which it was originally presented to the discoverers. All are resolved by a series of successive steps, determined beforehand and sufficiently numerous, so that no one step, so to speak, could not be understood by a suitably guided student.

What do we mean then by these last remarks? If the segmentation has been sufficient, it is possible that certain parts of the solution could be found spontaneously by good students. It would be unwise, in every case, to ask of all the students the same effort, and it seems to me to be entirely useless to restrict ourselves to that which can be resolved in this way.

How then can our questions give rise to those things which would not appear without the questions?

Should they be a form of dissimulation in which we indicate the solution, in such a way that our interlocutor believes that he has found the solution while all had been suggested to him without his having doubted it? This manner of guiding is that of Socrates himself (at least if in the absence of an original account, we can rely on a disciple—was he called Plato?—to render faithfully the thought of the master).

Mr. Durand criticizes this method; it believes it has attained its objective when it has misled the student into thinking he has obtained results, as opposed to those he is capable of obtaining.

On this last point, we can answer, with Mr. Tannery, that the assignments are there to place the student face to face with his weaknesses and that there is a first and valuable way of applying the heuristic method.

On the other hand is it certain that the illusion which Mr. Durand justifiably mistrusts, is specific to this method? If we wish to search for it wherever it exists, we should find it without difficulty in current instruction, thanks to the fragmentation to which we have alluded and which makes the mind, which is so important and so fruitful, vanish from the essence of the difficulty. Moreover if there is anything that gives birth to illusions that are especially deceptive, it is precisely lectures which are too well prepared, as we mentioned earlier. This arises at the moment when it is necessary to recall and rediscover these elegant constructions. At first sight they are seductive; others less perfect take their retribution the next day: in particular proofs obtained heuristically.

Moreover is it necessary that they be truly heuristic? That is never the case, as we have said, for the authentic Socratic method. The weakness of the initiative which it stimulates has some value—this is to show constantly to the professor whether or not the preceding material has been understood. We are correct, on the whole, in seeing nothing but an expository method which is less direct, less clear, and less convenient, and we are justified in reproaching it, and reproaching it without leniency, for raising all the difficulties which intimidate us from using the heuristic method, since this method exhibits none or very small advantages.

Moreover, this deception, these questions so easily posed, and somewhat specious, do not seem to me to have been widely disseminated. Mr. Durand compares a student to a child which a magician has put on the stage in order to pull from his pockets, from his nose, from his ears a multitude of unexpected and unusual objects. Very well! It is essential that the magician divulge his trick, that the teacher explains how these questions are carried to the end.

In reality, the mistake is to believe that there is a trick, a slight of hand of some sort, that we should imitate an examining magistrate—a dishonest examining magistrate—trying to get a confession from a guilty one. The nature of our questions, in my opinion, is something else and very simple: they should be those that the student asks himself, those which, learning from this experience, he asks himself a second time.

What these questions should be, how one should understand how to ask them,—not only the teacher but the student as well, that is what I have tried to outline elsewhere.

I tried, in other words to discern the rules, that all mathematicians unconsciously follow when they reason, at least the main ones; those rules that we find essentially the same in most of the examples. They are on the other hand common sense rules—truisms so to speak. In summary they amount to correctly posing the main problem.

Should we say that that always suffices to solve the problem? Evidently not—no one has the presumption to reduce science to a machine. Once we have understood how the question is posed, there remains nevertheless in general, something more to be found.

Of these two elements of the solution, the application of necessary logical rules and so-called invention, which of these is most often missing in students? Everything points to believing it is the second; experience shows that it is the first, that which it would seem should not be essential.

Take the example of a circle; the student will not be unaware of the definition of a circle; he will say it without hesitation if you ask him. But of his own accord, he does not ask, nor does he think, that this is the time to recall the definition. He neither thinks of asking himself what is the hypothesis of the theorem he is trying to prove, nor does he think of using it. It is one of the causes, or an analogous one, which gives rise to his impediment in understanding the proof. When we are compelled to substitute the definition to that which is defined, to use the entire hypothesis, when too, we have forced him to convince himself that every one of the transformations which he applied to the question does not change its true significance, we can determine that the help which is thus given, is, most of the time, the only one he needed. (One has only to question the students of the faculty to be aware of the ignorance which currently subsists on this matter after years of mathematical studies.)

Teaching thus understood, although heuristic, remains didactic in a certain sense as we see: it teaches the rules of the method instead of the results. However, naturally, this is not an a priori exposition; it will be practical, and will intervene to the extent that occasions arise for using it.

We see in this way how we may achieve this double ideal, seemingly contradictory, of a method that neither abandons the pupil to his own devices nor provokes him directly or indirectly about that which he should say.

We see that we are far from the criticism of dissimulation that earlier we feared we should incur.

Not only should we not mislead our students but we should show them carefully that which they had missed. We should not fear, under these circumstances, the serious danger with which Mr. Durand threatens us and that had especially struck Mr. Tannery: this sort of initial hesitation, thanks to which the mind always awaits, the initial external impulse which puts it into motion. The rules of which we have spoken have precisely the object of providing this impulse. It is up to us to force the student to make use of it and to make him aware that he is at fault for not applying these rules.

To arrive at this end, is it necessary to treat all these question heuristically as we had earlier assumed? Evidently not: it is neither desirable nor indispensable. We should apply the heuristic method not to the point of drowning ourselves, but nevertheless sufficiently frequently, not only to have it understood but to realize that we could have used it always if we had wished to do so.

We leave aside several truly difficult issues, where the difficulty cannot be broken into segments, where one must intervene in a manner other than by the general rules of which we have spoken, combined with the intuition which we can naturally get from the student. These, to the extent they exist, are the exception; the other difficulties are so numerous that we must choose among them. We can also discard, in particular, not only solutions which are more or less artificial—it is clear that in the first place one should not teach them, as far as possible, neither by the heuristic method nor otherwise—proofs which are very perfect, very systematic, those in which we are too frequently forced to conceal the essential idea by the precision of the details—always fossilized ideas. The theorem of projections was designed to be taught without an instructor. The same is true of the method of isoperimetry, if it had not, fortunately, disappeared from the curriculum; the same is not true of the method of perimeters. Not only does the last method not contain anything that cannot be "rediscovered" but the student can and should become aware that, except possibly for one point, he does not need to learn it in order to know it.

The experience will so easily show to each of them the appropriate level, that nothing will impede their being able to operate in a broad manner.

Nothing inhibits us from introducing the heuristic method in a unique form. It can be substituted for didactic learning; but it can be usefully used beforehand, the teacher summarizing—or sometimes having summarized, in a concise form, the reasonings which were found by the class, if he decides that order and clarity require them.

Here I am nearly in agreement with Mr. Durand who—I am not reluctant to believe it—sometimes also uses the heuristic method and had not expected that we speak of it. But we also see on which issue I dissociate myself from him. I do not believe that he agrees, as I insist, to apply the method sufficiently often to show that we can, and even that we must, apply it almost always.

However it is precisely that which I consider essential. The important point is not to arrive at the stage of using it in class, but to use it after class, when the student is left to fend for himself—pardon me for this tautology which is nothing but obvious—in the solution of problems. We therefore understand that, as Mr. Tannery remarked, the analysis of these same problems is particularly important if we take this point of view especially, and above all, in the study and revision of curricula, from one lesson to another. It is there, in my view, that we should not neglect using the heuristic method as continuously as possible. The student should, in principle, become conditioned in not knowing any other.

It is that, if I am not mistaken, that we should understand by "learning to learn." Either this device—which we do not apply as often as we speak about it—has nothing to say, or it means that graduating from school, youths will be enabled to use it when they replace the professors who abandon it. I do not know if the heuristic method will enable us to realize this ideal but I believe that it alone, suitably applied in the area which concerns us, can claim to work efficaciously.

At the age of eleven, I began Euclid. This was one of the great events in my life, as dazzling as first love.
—Bertrand Russell (1872–1970)

Mathematical Proof

Godfrey Harold Hardy

Godfrey Harold Hardy (1877–1947) was one of the towering figures in British mathematics during the first half of the twentieth century. His parents were both educators but coming from modest economic and social classes, they were deprived of a university education. This, fortunately, was not a deterrent and they supervised the enlightened education of their children—Harold and his younger sister Gertrude.

Hardy attended Winchester, one of the prestigious "public" (i.e., private) schools in England, and based on his outstanding scholastic performance, he went from there to Trinity College, Cambridge, from which he graduated in 1898. He was elected a fellow in 1900. From 1900 till 1911, he demonstrated his mathematical creativity but his most influential work of that period was a book published in 1908 entitled *A Course of Pure Mathematics*. This book which introduced readers to "rigorous" analysis transformed university teaching of mathematics throughout Britain and, indeed, elsewhere.

In 1911 he embarked on a lifelong collaboration with J. E. Littlewood—a collaboration that was extremely fruitful and consequential. The collaboration lasted 35 years. Meanwhile in 1913, a correspondence with the Indian mathematician S. Ramanujan resulted in a second major collaboration, which was short-lived as a result of the untimely death of Ramanujan.

Hardy was an avid cricketer and an engaging conversationalist discoursing on a wide variety of subjects. Politically, we should describe him today as a "liberal"; his family background may have been the basis for his mild disdain for the English bourgeoisie. He was strongly opposed to the First World War but years later he vigorously defended Bertrand Russell's dismissal in a book called *Bertrand Russell and Trinity*. In the preface to the book mentioned below, C. P. Snow explained that although Hardy was vehemently antiwar, "he could not find a satisfactory basis for conscientious objection."

On the personal side it is said that he was pleasant though he harbored some biases and his character was marked by some idiosyncrasies. Generous in his praise of others, he was very good company according to Snow. These qualities serve to complete a picture of a brilliant if somewhat eccentric scholar who had a touch of panache.

Many of his mathematical contributions have an enduring quality and many results bear his name—including one which has application to genetics. One of his most popular works, however, is a sort of autobiography entitled *A Mathematician's Apology*, written in 1940. It was very widely read and it seems likely that Hardy is not "apologizing" for being a mathematician but is rather using the word in the more classical sense of "justification." The book is still extensively distributed and read and is a rich source of Hardy's personal views, some piquant; many of these have become mathematical household quotations.

Hardy suffered a heart attack and his last days were sorrowful; he died in 1947.

Editor's Preface

This interesting essay is Hardy's response to what were widespread contemporary deliberations concerning the epistemology and ontology of mathematics and its concepts. By epistemology is meant the origin and nature of knowledge while ontology refers to a description of its existence or the basic characteristics of its reality.

The essay is relevant to our times since the current debate about the nature and content of mathematics is lively and extensive. The discussion is in the context of the philosophy of mathematics, a discipline dating from ancient Grecian times..

Philosophy of mathematics encompasses four principal schools of thought. There are other points of view but we shall limit the discussion to these most common ones. A brief description will not do adequate justice, nor will it give any deep perception of the differing philosophies, but it will give the reader an idea of the different points of view and we hope will enable the reader to appreciate the issues with which Hardy deals.

(1) Logicists. This school claims that mathematics is a part of logic, that the theorems and definitions of mathematics may be expressed in the language of logic and set theory. This school was founded by Gottlob Frege and reached its maturity with *Principia Mathematica*, the work of Bertrand Russell and Alfred North Whitehead.

Unfortunately difficulties arise when we come to deal with infinite sets. These give rise to contradictions, or as Hardy calls them antinomies. Perhaps the best-known example is that of Russell's paradox, viz., "the set S of all sets which are not members of themselves." Although Russell offered a resolution of this difficulty with his theory of types, many scholars remained dissatisfied with the logicists' point of view. Among the dissatisfied were the

(2) Intuitionists. This group is very suspicious of the use of infinite sets and of mathematical entities that are not explicitly constructable or of mathematical quantities that are not explicitly determinable. They are accordingly led to reject the law of the excluded middle. An example due to John von Neumann illustrates the disquiet expressed by these advocates. The example is a piecewise linear function the location of whose root depends on the validity, or lack of it, of Goldbach's conjecture that every even integer is the sum of two primes. Goldbach's conjecture remains in the realm of conjecture.

(3) Formalists. This point of view was strongly advocated by David Hilbert—see the essay by that author.

Formalists claim that the elements of mathematics need have no intrinsic meaning. The entire process is a formal game in which the symbols are manipulated according to the rules of logic. Two statements are equal if each is provable from the other. Hilbert's first example was the axiomatization of Euclidean geometry in which the elementary concepts of point, line, and plane are independent of any physical reality and may be replaced by three arbitrary words.

Once again this point of view encounters difficulties in dealing with infinite sets.

(4) Platonists. Platonists believe that mathematics exists in some ideal universe and the task of the mathematician is to discover its truths. It says nothing about the methods and techniques of discovery nor about the nature of the truths. A person may be a Platonist and also espouse some aspects of one of the above schools.

A mathematician who has not had much experience dealing with philosophical concepts and those of logic may be bedazzled by the intricacy of the underlying semantics. For example, how do formalists differ from logicists? Moreover in delving into these matters, one quickly encounters the fundamental logical issues of completeness and decidability. In short, there is little doubt that in probing the ontological nature of a system of mathematics which incorporates the infinite, one encounters difficulties that have not been resolved to the satisfaction of all.

It would appear that Hardy is basically a Platonist but we shall let him speak for himself.

Note. The editor has abridged the original article—primarily to make it more accessible to the non-specialist. Hardy's point of view has not been altered.

1. I have chosen a subject for this lecture, after much hesitation, not from technical mathematics but from the doubtful ground disputed by mathematics, logic and philosophy; and I have done this deliberately, knowing that I shall be setting myself a task for which I have not sufficient qualifications. I have been influenced by three different motives. In the first place, the exercise will be good for me, since it will force me to think seriously about questions which a professional mathematician like myself is apt to neglect. Secondly, it is difficult to find a branch of pure mathematics suitable for popular exposition in an hour. Finally if, in a desperate attempt to be interesting, I lose myself in discussions where I am admittedly an amateur, then, whoever I may offend, I should certainly not have offended the founder of this lectureship and the Rouse Ball chair.

I do not regret my choice, but I am bound in self-defense to begin with a double apology. The first is to any real mathematical logicians who may be present. I am myself a professional pure mathematician in the narrow sense, and, in my own subject, quite as intolerant of amateurs as a self-respecting professional should be. I have therefore no difficulty in understanding that mathematical logic also is a subject for professionals; that it demands a detailed knowledge which I do not possess and, so long as I am active in my proper sphere, have hardly leisure to acquire; and that I am certain to be guilty of all sorts of confusions which would be impossible to a properly qualified logician. Indeed there is only one thought which gives me courage to proceed, and that is that I may be concerned less with strictly logical questions than with questions of general philosophy. However treacherous a ground mathematical logic, strictly interpreted, may be for an amateur, philosophy proper is a subject, on the one hand so hopelessly obscure, on the other so astonishingly elementary, that there knowledge hardly counts. If only a question be sufficiently fundamental, the arguments for any answer must be correspondingly crude and simple, and all men may meet to discuss it on more or less equal terms.

My second apology must be addressed to those mathematicians who dislike all discussions savoring of philosophy. But if I apologize to them, it is perhaps with less sincerity. I feel that this distaste is usually based on no better foundation than an unreasoning shrinking from anything unfamiliar, the distaste of the pragmatist for truth, of the engineer for mathematics, of the pavilion critic at Lords for the in-swinger and the two-eyed stance. It is reasonable to ask an audience like this to put aside this dislike of the fundamental for its own sake.

You must also remember that ordinary mathematics has a good deal at stake in some of these recent controversies. These controversies have seemed to threaten methods which we have used with confidence for nearly one hundred years. There are familiar elementary theorems the truth of which is simply denied by the 'intuitionist' school of logicians. There are also theorems of an apparently much less abstract or suspicious type, theorems for example in the theory of numbers, the only known proofs of which depend, in appearance at any rate, on principles which they reject.

2. It may not be possible to distinguish precisely between mathematics, mathematical logic, and philosophy, as the words are currently used. We can, however, by considering a few typical problems, recognize roughly the disputed tracts across which the boundaries must be drawn.

(i) *Is Goldbach's Theorem true?* Is any even number the sum of two primes? This is a strictly mathematical question to which all questions of logic or philosophy seem irrelevant.

(ii) *Is the cardinal number of the continuum the same as that of Cantor's second number class?* This again appears to be a mathematical question; one would suppose that, if a proof were found, its kernel would lie in some sharp and characteristically mathematical idea. But the question lies much nearer to the borderline of logic, and a mathematician interested in the problem is likely to hold logical and even philosophical views of his own.

(iii) *What is the best system of primitives for the logic of propositions?* This is a question of mathematical logic in the strict professional sense.

(iv) *What is a proposition, and what is meant by saying that it is true?* This, finally, is a problem of simple philosophy.

It is often said that mathematics can be fitted on to any philosophy, and up to a point it is obviously true. Relativity does not (whatever Eddington may say) compel us to be idealists. The theory of numbers does not commit us to any particular view of the nature of truth. However that may be, there is no doubt that mathematics does create very strong philosophical prejudices, and that the tests which a philosophy must satisfy before a mathematician will look at it are likely to be very different from those imposed by a biologist or a theologian. I am sure that my own philosophical prejudices are as strong as my philosophical knowledge is scanty.

One may divide philosophies into *sympathetic* and *unsympathetic*, those in which we should like to believe and those which we instinctively hate, and into *tenable* and *untenable*, those in which it is possible to believe and those in which it is not. To me, for example, and I imagine to most mathematicians, Behaviorism and Pragmatism are both unsympathetic and untenable. The Cambridge New Realism, in its cruder forms, is very sympathetic, but I am afraid that, in the forms in which I like it best, it may be hardly tenable. 'Thin' philosophies, if I may adopt the expressive classification of William James, are generally sympathetic to me, and 'thick' ones unsympathetic. The problem is to find a philosophy which is both sympathetic and tenable; it is not reasonable to hope for any higher degree of assurance.

3. The crucial test of a philosophy, for a mathematician, is that it should give some sort of rational account of *propositions* and of *proof*. A mathematical theorem is a proposition; a mathematical proof is clearly in some sense a collection or pattern of propositions. It is plain then that if I ask what are, to a mathematician, the most obvious characteristics of a mathematical theorem or a mathematical proof, I am inviting philosophical discussion of the most fundamental kind. I wish to begin, however, by being as unsophisticated as I can, and I will therefore try to sketch what seems to be the view of mathematical common sense, the sort of view natural to a man who does not profess to be a logician but has spent his life in the search for mathematical truth. It is after all the misapprehensions of such a man that a logician may find the least fundamentally unreasonable and the least hopeless to remove.

I will begin then by enumerating some rough criteria which I think that a philosophy must satisfy if it is to be at all sympathetic to a working mathematician. I know too well how probable it is that just the most sympathetic philosophies will prove untenable.

(i) It seems to me that no philosophy can possibly be sympathetic to a mathematician which does not admit, in one manner or another, the immutable and unconditional validity of mathematical truth. Mathematical theorems are true or false; their truth or falsity is absolute and independent of our knowledge of them. In some sense, mathematical truth is part of objective reality.

'Any number is the sum of 4 squares' ; 'any number is the sum of 3 squares' 'any even number is the sum of 2 primes.' These are not convenient working hypotheses, or half-truths about the Absolute, or collections of marks on paper, or classes of noises summarizing reactions of laryngeal glands. They are, in one sense or another, however elusive and sophisticated that sense may be, theorems concerning reality, of which the first is true, the second is false, and the third is either true or false, though which we do not know. They are not creations of our minds; Lagrange discovered the first in 1774; when he discovered it he discovered *something*; and to that something Lagrange, and the year 1774, are equally indifferent.

(ii) When we know a mathematical theorem, there is something, some object, which we know; when we believe one, there is something which we believe; and this is so equally whether what we believe is true or false.

It is obvious that by this time we have escaped only too successfully from the domain of platitude and triviality. We have done no more than to make explicit a few of the instinctive prejudices of the 'mathematician in the street'. Yet with our first demand we have antagonized at least two-thirds of the philosophers in the world; and with the second we have reduced our first indiscretion to entire insignificance, since we have committed ourselves, in one form or another, to the objective reality of propositions, a doctrine rejected, I believe, not only by all philosophers, but also by all three of the current schools of mathematical logic.

(iii) In spite of this I am going farther, and in a direction relevant to the recent controversies concerning 'transfinite' mathematics to which

I shall return later. Mathematicians have always resented attempts by philosophers or logicians to lay down dogmas imposing limitations on mathematical truth or thought. And I am sure that the vast majority of mathematicians will rebel against the doctrine—even if it is supported by some of themselves including mathematicians so celebrated as Hilbert and Weyl—that it is only the so-called 'finite' theorems of mathematics which possess a real significance. That 'the finite cannot understand the infinite' should surely be a theological and not a mathematical war-cry.

No one disputes that there are infinite processes which appear to be prohibited to us by the facts of the physical world. It is true, as Hilbert says, that no mathematician has completed an infinity of syllogisms. It is equally true that there is no mathematician who has never drunk a glass of water, and, so far as I can see, one of these facts has neither more nor less logical importance than the other. There is no more logical reason why a mathematician should not prove an infinity of theorems in this world than why he should not (as he has been so often encouraged to hope) emit an infinite sequence of musical notes in the next.

The history of mathematics shows conclusively that mathematicians do not evacuate permanently ground which they have once conquered. There have been many temporary retirements and shortenings of the line, but never a general retreat on a broad front. We may be confident that, whatever the precise issue of current controversies, there will be no general surrender of the ground which Weierstrass and his followers have won. 'No one', as Hilbert says himself, 'shall chase us from the paradise that Cantor has created'.The worst that can happen to us is that we shall have to be a little more particular about our clothes.

4. Such then are the presuppositions and prejudices with which a working mathematician is likely to approach philosophical or logical systems. How far are they satisfied by the existing schools of mathematical logic? There are three such schools, the logisticians (represented at present by Whitehead, Russell, Wittgenstein, and Ramsey), the finitists or intuitionists (Brouwer and Weyl), and the formalists (Hilbert and his pupils). I am primarily interested at the moment in the formalist school, first because it is perhaps the natural instinct of a mathematician (when it does not conflict with stronger desires) to be as formalistic as he can, secondly because I am sure that much too little attention has been paid to formalism in England, and finally because of the title of my lecture and because Hilbert's logic is above everything an explicit theory of mathematical proof. I must begin by a rapid summary of the most striking differences between these schools and of the difficulties which have brought them into existence.

5. (i) I shall refer to the logisticians generally under the short title of 'Russell.' It is necessary to say that by 'Russell' I mean the Russell of *Principia Mathematica. Principia Mathematica* is not a treatise on philosophy, but it has a philosophical background, with which I am in general sympathy.

To Russell, then, logic and mathematics are substantial sciences which in some way give us information concerning the form and structure of reality. Mathematical theorems have *meanings*, which we can understand directly, and this is just what is important about them. In this, I may observe, Russell and the so-called 'intuitionists' are in complete agreement; and (since it is something of this sort which seems to me the natural implication of the word) I should prefer to avoid the use of 'intuitionism' as distinguishing one school from the other.

Mathematics is to Russell, up to a certain point at any rate, a branch of logic. It is concerned with particular kinds of assertions about reality, with particular logical concepts, propositions, classes, relations, and so forth. The propositions of logic and mathematics share certain general characteristics, in particular complete generality, though this is not an adequate description of them. There is no particular reason that I can see why any of this should be distasteful to us as mathematicians. It does not seem to conflict with the criteria which I suggested a moment ago; it seems likely at first sight even to indulge our desire for real propositions, though here we are ultimately disappointed.

There are certain definite points at which Russell's attempted reduction of mathematics to logic fails. In this, of course, there is nothing likely to astonish an unsophisticated mathematician. That mathematics should follow naturally, up to a point, from purely logical premisses, premisses to whose simplicity and self-evidence no one can reasonably take exception, when proper allowance is made for the element of sophistication inevitable in a highly complex structure; but that it should then prove necessary to import fresh raw material and add new assumptions—all this is only what a mathematician might expect. In particular, I think that this is true of two of the three 'non-logical' axioms necessary in Russell's scheme; the Axiom of Infinity, that the universe contains an infinity of individuals, and the Multiplicative Axiom or Axiom of Zermelo, which is very famous but required only in particular theorems which might conceivably be discarded, and which I need not stop to explain, since I shall not refer to it further.

6. The theory of aggregates, in the classical form of Cantor and Dedekind, leads to certain antinomies, of which the most famous is Russell's paradox of the class of classes which are not members of themselves, a concept which may be shown to lead at once to flat contradiction. Russell met the difficulty by his Theory of Types.

The theory of types has, however, very unfortunate mathematical consequences, since it appears to destroy some of the most fundamental theorems of analysis. The typical theorem is the theorem that any aggregate of numbers has an upper bound, a theorem which is substantially the same as what, in my *Pure Mathematics*, is called 'Dedekind's Theorem'.

7. I pass to the finitists, Brouwer and Weyl, and I shall dismiss them very shortly. Much as I admire the contributions of Brouwer and Weyl to constructive mathematics, I find their contribution to logic singularly unsympathetic. Finitism rejects, first, all attempts to push the analysis of mathematics beyond a certain point, and for this I see no sort of justification. I have no particular desire to be committed to the extreme Russellian doctrine, that all mathematics is logic and that mathematics has no fundamentals of its own.

This, however, is a minor point for a mathematician. What is much more serious to a mathematician is that the mathematical consequences of finitism involve rejection not of particular isolated outworks of mathematics but of integral regions of ordinary analysis. It is no use trying to deny that the finitists have the better of the argument up to a point; the parts of analysis which they admit are unquestionably, at present, in a more secure position than the rest; and so long as finitism merely insists on this its position is unassailable. I cannot believe that mathematicians generally will be so ready to accept a check as final, so anxious to find metaphysical reasons for supposing that the prettiest path is that which passes on the side of the hedge away from the bull.

8. I go on then to consider the logic of Hilbert and his school; and here I find it very necessary to distinguish between Hilbert the philosopher and Hilbert the mathematician. I dislike Hilbert's philosophy quite as much as I dislike that of Brouwer and Weyl, but I see no reason for supposing that the importance of his logic depends in any way on his philosophy.

I am sure that the Hilbert logic has been unreasonably neglected by English logicians. 'The formal school', says Ramsey, 'have concentrated on the propositions of mathematics, which they have pronounced to be meaningless formulæ to be manipulated according to certain rules, and mathematical knowledge they hold to consist in knowing what formulæ can be derived from what others consistently with the rules. Such being the propositions of mathematics, the account of its concepts, for example the number 2, immediately follows: "2" is a meaningless mark occurring in these meaningless formulæ. But, whatever may be thought of this as an account of mathematical propositions, it is obviously hopeless as a theory of mathematical concepts; for these occur not only in mathematical propositions, but also in those of everyday life. Thus "2"occurs not merely in " 2 + 2 = 4," but also in "it is 2 miles to the station," which is not a meaningless formula but a significant proposition, in which "2" cannot conceivably be a meaningless mark. Nor can there be any doubt that "2" is used in the same sense in the two cases, for we can use "2 + 2 = 4" to infer from "it is 2 miles to the station and 2 miles on to the Gogs" to "it is 4 miles to the Gogs *via* the station," so that these ordinary meanings of' 2" and "4" are clearly involved in "2 + 2 = 4."

Let me say at once that this argument seems to me to be unanswerable and that, if I thought that this really was the beginning and the end of formalism, I should agree with Ramsey's rather contemptuous rejection of it. But is it really credible that this is a fair account of Hilbert's view, the view of the man who has probably added to the structure of significant mathematics a richer and more beautiful aggregate of theorems than any other mathematician of his time? I can believe that Hilbert's philosophy is as inadequate as you please, but not that an ambitious mathematical theory which he has elaborated is trivial or ridiculous. It is impossible to suppose that Hilbert denies the significance and reality of mathematical concepts, and we have the best of reasons for refusing to believe it: "the axioms and demonstrable theorems," he says himself, "which arise in our formalistic game, are the images of the ideas which form the subject-matter of the ordinary mathematics."

9. It is no doubt this philosophical outlook, and this consequent insistence on the importance of the physical mark or sign, that inspire Hilbert's finitism, which appears at first sight as extreme as that of Brouwer and Weyl themselves. I naturally find this attitude very disappointing; it seems to me that formalism is bound to die for want of air within the narrow confines of a finitistic system. But on the face of it Hilbert is entirely uncompromising: "there is no infinite anywhere in reality," he says, and again, "is it not clear that, when we think we can recognize the reality of the infinite in any sense, we are merely allowing ourselves to be deceived by the enormity of the largeness or smallness which confronts us everywhere...."

Hilbert says that 'infinite theorems,' theorems such as 'there are infinitely many primes,' are not genuine propositions but 'ideal' propositions. I am not at all sure what he means by an 'ideal proposition,' but I suppose that one thing at any rate that he would say (if he used Russell's language) is that the infinite is essentially incomplete. We know that mathematics is full of 'incomplete symbols,' symbols which have no meaning in themselves, though larger collections of symbols of which they are parts have perfectly definite meanings. There is, in the classical analysis, no number ∞ standing on all fours with e or π; there is a sharp contrast here between the infinite of analysis and the infinite of geometry, in which 'the line at infinity', say $z = 0$, is on just the same footing as any other line.

It is one of Russell's admitted achievements to have recognised in a precise and explicit manner the immense importance of 'incomplete symbolism' in logic and philosophy also, and so to have shown how widely the correct analysis of a proposition may diverge from the analysis of unreflecting common sense.

I am not suggesting that Hilbert would accept the statement that the infinite is incomplete as an adequate account of his attitude towards it. No doubt he would want to go very much further. I have inserted this explanation merely (1) because I shall need it later and (2) because rival views about the infinite are apt to differ more violently in expression than reality, and the notion of an incomplete symbol might in some cases be a basis for a reconciliation between them. I have the less hope that it would do so in this case because Hilbert uses, as instances in support of his thesis that all ' infinite theorems' are in some sense ' ideal theorems', such divergent illustrations as (a) the infinite of analysis, (b) the infinite of geometry, and (c) the ideal numbers of higher arithmetic, and it seems to me quite impossible to regard all these as inspired by the same logical motive, the first representing a *purification* of mathematics by an agreement to regard certain notions as 'incomplete,' the others an *enlargement* of it by the introduction of new elements as 'complete' as those which they generalize.

10. It is time, however, to proceed to some description of Hilbert's system, and I do this in language based upon that of v. Neumann, a pupil of Hilbert's whose statement I find sharper and more sympathetic than Hilbert's own.

(i) Hilbert's logic is a theory of proof. Its object is to provide a system of formal axioms for logic and mathematics, and a formal theory of logical and mathematical proof, which (a) is sufficiently comprehensive to generate the whole of recognized mathematics, and (b) can be proved to be consistent. The system of *Principia Mathematica* fulfils the first but not the second criterion.

(ii) If we can do this, we shall be troubled by antinomies no more. But for this end the whole existing apparatus of axioms, proofs and theorems must first be formalized strictly, so that to every mathematical theorem a formula will correspond. The structure of the formal system will of course be *suggested* by the current logic and mathematics. Every formula will *seem* to have a meaning, a meaning which we must afterwards forget.

(iii) For example, we have the 'logical' formula

$$a \Rightarrow (b \Rightarrow a)$$

This is suggested by an obvious 'logical truth' that a true proposition is a consequence of any hypothesis. This formula is an "axiom" , which means simply that it is one of the formulae with which we start.

(iv) Let us observe in passing that there are far more axioms in Hilbert's scheme than in such a scheme as that of *Principia Mathematica*, and no *definitions* in the sense of *Principia Mathematica.*

11. Similarly there is the Hilbert mathematics on the one hand, and what Hilbert calls 'metamathematics' on the other, the metamathematics being the aggregate of theorems *about* the mathematics; and of course it is the metamathematics which is the exciting subject and affords the real justification for our interest in this particular sort of mathematics. Suppose, for example, that we could find a finite system of rules which enabled us to say whether any given formula was demonstrable or not. This system would embody a theorem of metamathematics. There is of course no such theorem, and this is very fortunate, since if there were we should have a mechanical set of rules for the solution of all mathematical problems, and our activities as mathematicians would come to an end.

Such a theorem is not to be expected or desired, but there are metamathematical theorems of a different kind which it is entirely reasonable to expect and which it is in fact Hilbert's dominating aim to prove. These are the negative theorems of the kind which I illustrated a moment ago; they assert, for example, in chess, that two knights cannot mate, or that some other combination of the pieces is impossible, in mathematics that certain theorems cannot be demonstrated, that certain combinations of symbols cannot occur. In particular we may hope (and it is this hope that has inspired the whole construction of the logic) to show the impossibility of the combination

$$a \text{ and } (\text{not } a)$$

12. It is now time for me to interpolate a remark which gives the justification for the title of my lecture. It is obvious that to Hilbert *proof* means two quite different things. I have tried to anticipate the point in my choice of words: we fortunately have two words, *proof* and *demonstration.*

'*Proof*' has always meant at least two different things, even in ordinary mathematics. We distinguish vaguely and halfheartedly; in the Hilbert logic, the distinction becomes absolutely sharp and clear. First, there is the *formal*, *mathematical*, *official* proof, the proof inside the system, the pattern (A), what I called the *demonstration.* These inside official proofs are, in the mathematics, the actual formulæ or patterns, in the metamathematics the subject matter for discussion.

Second there are the proofs of the theorems of the metamathematics, the proof that two knights cannot mate. These are informal, unofficial, significant proofs, in which we reflect on the meaning of every step. The structure of these proofs is not dictated by our formal rules; in making them we are guided, as in ordinary life, by 'intuition' and common sense. 'Prof. Hardy will lecture at 12:00 to-day, because it says so in the Reporter, and because statements in the Reporter are always true.'

You must not imagine that the unofficial, metamathematical, nonformal, intuitionist proof is in any sense slacker or less 'rigorous' than the formal mathematical proof. The subject matter is abstract and complicated, and every step has to be scrutinised with the utmost care.

13. At this point I should like to leave the Hilbert logic for a moment, and make a few general remarks about mathematical proof as we working mathematicians are familiar with it. It is generally held that mathematicians differ from other people in proving things, and that their proofs are in some sense grounds for their beliefs. Dedekind said that 'what is provable, ought not to be believed without proof'; and it is undeniable that a decent touch of scepticism has generally (and no doubt rightly) been regarded as some indication of a superior mind.

But if we ask ourselves why we believe particular mathematical theorems, it becomes obvious at once that there are very great differences. I believe the Prime Number Theorem because of de la Valleé-Poussin's proof of it, but I do not believe that 2 + 2 = 4 because of the proof in *Principia Mathematica.* It is a truism to any mathematician that the 'obviousness' of a conclusion need not necessarily affect the interest of a proof.

I have myself always thought of a mathematician as in the first instance an *observer*, a man who gazes at a distant range of mountains and notes down his observations. His object is simply to distinguish clearly and notify to others as many different peaks as he can.

The analogy is a rough one, but I am sure that it is not altogether misleading. If we were to push it to its extreme we should be led to a rather paradoxical conclusion; that there is, strictly, no such thing as mathematical proof; that we can, in the last analysis, do nothing but *point*; that proofs are what Littlewood and I call *gas*, rhetorical flourishes designed to affect psychology, pictures on the board in the lecture, devices to stimulate the imagination of pupils. This is plainly not the whole truth, but there is a good deal in it. The image gives us a genuine approximation to the processes of mathematical pedagogy on the one hand and of mathematical discovery on the other; it is only the very unsophisticated outsider who imagines that mathematicians make discoveries by turning the handle of some miraculous machine. Finally the image gives us at any rate a crude picture of Hilbert's metamathematical proof, the sort of proof which is a *ground* for its conclusion and whose object is to *convince.*

On the other hand it is not disputed that mathematics is full of proofs, of undeniable interest and importance, whose purpose is not in the least to secure conviction. Our interest in these proofs depends on their formal and æsthetic properties. Our object is *both* to exhibit the pattern and to obtain assent. We cannot exhibit the pattern completely, since it is far too elaborate; and we cannot be content with mere assent from a hearer blind to its beauty.

14. Let us return to the Hilbert logic. The very structure of the logic, its mere existence, are enough, I think, to prove two propositions of great importance. The first is that it is possible to establish the consistency of a system of axioms *internally*, that is to say by direct examination of its structure; and the second is that it is possible to prove a system consistent even when the axioms embody logical principles such as the law of contradiction itself. Each of these propositions has been disputed.

Consider for a moment the ordinary procedure of axiomatic geometry. In abstract geometry we consider unspecified systems of things, a class S of objects A, B, C,... which we call *points*, and sub-classes of these objects which we call *lines*. We make certain assumptions about these points and lines, which we call *axioms*, such as that there is a line which contains any given pair of points, that there is only one such line, and so on. To lay down a system of axioms in geometry is simply to limit the subject matter, to say that we propose to consider only objects of certain kinds. Thus, in a geometry which contains the two axioms I have mentioned, our 'points' might be the players in a tournament, and our 'lines' the opponents in a game, but the points and lines could not be undergraduates and colleges, because then the axioms would be untrue.

In a geometry we are not concerned with any *particular* meaning of 'point' or 'line.' We may say, if we like, that we are concerned with *all possible* meanings, or that we are not concerned with meanings at all.

Every geometry demands a *consistency theorem*, which is naturally not a theorem of the geometry. We have to prove that the axioms do not contradict one another. We produce an example, an 'interpretation', of the geometry, a set of objects which actually have the properties attributed by the axioms to our points and lines.

If we try to apply a similar process to arithmetic, we are met by a difficulty. There is, however, an obvious difficulty about the inevitable proof of consistency. When we wanted such a proof for a geometry, we could appeal to arithmetic; but there is nothing in ordinary mathematics which comes before arithmetic.

Finally, if we have established consistency in geometry and arithmetic, can we do so in logic, or in a subject which includes logic? It has been held, and I think by Russell, that we cannot, because our formulæ symbolize, among other things, the logical processes which we use in examining it, because the rules of the game are required in forming the judgement that what purports to be an instance of the game really is one. Other logicians, with whom here I agree, have held that this is a misunderstanding, due to a failure to distinguish between the use of our symbolism inside and outside the formal system.

My own view is that even here the classical method, the method of instances, is available in principle, and that, in restricted subjects such as the logic of classes or of propositions, it can be and has been successfully carried through. If, however, we are as ambitious as Hilbert, so that our system is to cover the whole field of abstract thought, I imagine that the attempt to do what we want on these lines is hopeless. I cannot imagine where we could find an adequate image of so comprehensive a symbolism, except in the whole field of thought which it was actually constructed to symbolize. There remains only the 'internal' method followed by Hilbert, based on study of the formal properties of the rules themselves. Whatever we may think about the philosophical basis on which Hilbert has erected his system, and with whatever success he or his followers may pursue it, it seems to me unquestionable that this method is valid in principle, in mathematics in exactly the same sense as in chess. And in this case Hilbert is entirely justified in his claim that he has found a necessary condition for all systems of mathematical logic, and that 'even the assertions of intuitionism, however modest they may be, require first a certificate of authorization from this tribunal.'

15. My remarks up to this point have been mainly explanations of things which I think I understand. The rest of what I have to say amounts to little more than a confession of a series of perplexities.

The first question which you will naturally ask is this: granted that Hilbert's method is valid in principle, what has it *done*? How far has the proof of consistency progressed? *Does* it establish freedom from contradiction in a domain co-extensive with mathematics? So far as I know the answer is, up to the present, *No*. There has been very substantial progress, and consistency has been proved up to a point beyond the point up to which success might be expected to be easy. The region accounted for includes the mathematics of the finitists, and that part of *Principia Mathematica* which is independent of the Axiom of Reducibility; but this region does not cover analysis.

It would be very reasonable to ask me, as an analyst, to explain my own attitude towards this hiatus in the foundations of analysis, and I do not profess to be able to give any satisfactory answer. I could only say this: in the first place, I am no finitist; I believe that the analysis of the text-books is true. Secondly, Ramsey has advanced a solution, which he does not profess to regard as entirely satisfactory, but in which I can find a good deal of encouragement.

16. I will return for a moment in conclusion to the properly 'philosophical' question to which I referred at the beginning, about the reality or 'completeness' of propositions. I am entirely unable to exorcize my craving for real propositions, a weakness which is after all only natural in a mathematician, to whom mathematical theorems ought to be the first basic reality of life. But I can find no sort of encouragement wherever I turn.

Our first instinct is to suppose that a judgement, whether true or false, must be analyzable into a mind and an object in relation. In a sense this is admitted to be true by everybody; it is undisputed that there is something objective, what Russell and Wittgenstein call the 'proposition as fact,' which enters into any judgement. When we judge, we form a picture of the reality about which we are judging, a form of words, a set of marks or noises, which we suppose, rightly or wrongly, to afford an image of the facts. This is the 'proposition as fact'; the question is, what, if anything, is there more?

There is, however, something beside the picture or factual proposition, namely the proposition in the sense which is relevant to logic. What is relevant to logic is not the factual proposition but what is common to all the factual propositions that can be pictures of a given state of affairs. A proposition is thus in some sense, a *form*. The propositions of Hilbert's logic are also forms, but Wittgenstein's forms are more substantial than Hilbert's, since they contain what Russell and Wittgenstein call the 'logical constants', 'and', 'or', 'not' and so forth, whereas Hilbert's can hardly be said to 'contain' anything at all. These logical constants do not represent and are not represented, but are present in the proposition (that is to say the factual proposition) as in the fact. The proposition (that is to say here the logical proposition) is thus a form of logical constants, whereas Hilbert's propositions are, so to say, *pure* form.

I ask then, finally, whether there is anything in the proposition, as relevant to logic and as Wittgenstein seems to conceive it, which affords any justification for my belief in 'real' propositions, my invincible feeling that, if Littlewood and I both believe Goldbach's theorem, then there is something, and that the same something, in which we both believe, and that that same something will remain the same something when each of us is dead and when succeeding generations of more skillful mathematicians have proved our belief to be right or wrong. I hoped to find support for such a view, when I read that 'the essential in a proposition is that which is common to all propositions which can express the same sense' and that 'the proposition is the propositional sign in its projective relation to the world.' When I read further, both in the book itself and in what Russell says about it, I concluded that I had been deceived. I can find nothing, in Wittgenstein's theory that is common to all the ways in which I can say that something is true and is not common also to many of the ways in which I can say that it is false. So here I can find no support for my belief; and if not here, where am I likely to find it? Yet my last remark must be that I am still convinced that it is true.

Algebra and money are essentially levelers; the first intellectually, the second effectively.
— Simone Weil (1903–1943)

The Unity of Knowledge

Hermann Weyl

Hermann Weyl (1885–1955) was born in Elmshorn, a small town adjacent to Hamburg; he was the son of Ludwig and Anna Weyl. As a schoolboy he read his father's copy of Kant's *Critique of Pure Reason* and was enchanted with Kant's concept of the a priori nature of Euclidean geometry. His enthusiasm was quickly dispelled on reading Hilbert's work on the foundations of geometry.

At Göttingen he got a PhD under the direction of David Hilbert with a thesis on singular integral equations. After a brief period lecturing in Göttingen, he accepted a position at the Technische Hochschule in Switzerland, a position he held from 1913 until 1930 when he was called to the chair in Göttingen vacated by Hilbert. With the advent of the National Socialist Party the political conditions in Germany were intolerable to him, especially with the dismissal of "non-Aryan" mathematicians. Courted by many institutions, he finally accepted an appointment at the newly established Institute for Advanced Study in Princeton in 1933 and retired in 1952.

He married Helene Joseph in 1913; they had two sons. Helene shared Weyl's love of philosophy and her charm and wit made the Weyl home in Princeton a gathering place for lively intellectual discussions. She died unexpectedly in 1948. He married Ellen Bar in 1950 and together they spent their remaining years commuting between Zürich and Princeton.

Weyl is universally regarded as one of the most versatile and creative mathematicians of the first half of the 20th century. His interests ranged from number theory to the foundations of quantum mechanics. In every case he made profound contributions to the subject and he left an intellectual heritage whose importance has not diminished with the passage of time.

Although shy by nature, he was a very pleasant and engaging conversationalist who spoke and wrote eloquently and sometimes wittily. The renowned physicist John Wheeler, when once asked to characterize Weyl, used the word "nobility" as connoting a person of high moral character coupled with courage, generosity, honor and vision. Weyl was most eloquent in pointing out pinnacles of the past in the realm of knowledge and at the same time pointing out peaks yet to be conquered. He had a unique style of writing—not always transparent but ever full of grace as the following quotation from his book on *Philosophy and Natural Science* illustrates: "The objective world simply is, it does not happen. Only to the gaze of my consciousness, crawling upward along the life line of my body, does a section of this world come to life as a fleeting image in space which continuously changes in time."

In addition to his extensive fundamental research, he wrote many influential books. Among these were *Group Theory and Quantum Mechanics*, which laid a firm mathematical foundation for the subject. His book *Symmetry* was widely read and quoted. His discourse on *Philosophy of Mathematics and the Natural Sciences* went through several editions and his book on algebraic number theory is cherished by specialists in that field. In addition, his work on the idea of a Riemann surface was a welcome account of what was then regarded as an exotic concept.

He was very much at home in the celebrated works of literature, not only the German writers—Goethe, Rilke, Mann, among others on which he was fed as a student in his native Germany, but the works too of Shakespeare, Coleridge, T.S. Eliot as well as others.

To illustrate his genial sense of humor, he made an amusing comment on the *formula* for determining federal income tax liability. It is so archaic, he said, that it belongs to the mathematics of the 13th C.!

He was in great demand as a speaker for special occasions. The essay below was one such occasion. He died, we would now say prematurely, in 1955 at the age of 70.

Editor's Preface

The essay below is one expression of Weyl's lifelong interest in philosophy. His first exposure to philosophy was as a youth when he read Kant's *Critique of Pure Reason* and came under its spell. When however, as a mathematics student, he read Hilbert's axiomatization of Euclidean Geometry he became enchanted with mathematics and welcomed the release of Euclid from the restraints of Kant's a priori point of view.

Weyl soon became familiar with the work of the German philosophers—Hegel, Heidegger and Husserl—especially the last who had written on logic, among other things. Throughout his life Weyl continued his efforts to integrate philosophy with scientific endeavors.

The search for knowledge and human understanding is a yearning in humans that dates from the earliest times and was one of the burning issues in Greek philosophy. Over the millennia, idealism, materialism, epistemology, phenomenology and other philosophical constructs had, as one of their goals, a deeper penetration of this issue. Indeed Weyl has commented that this question is one of our most essential intellectual challenges. (The figures in the caves of Lascaux are an expression of this same challenge!)

When an individual not well versed in the philosophic tradition, encounters a philosophic discourse, he or she does so with some fear and trembling. The underlying reason is that the free flow of philosophical ideas is not always compatible with the objective clarity and apparent determinacy of the natural sciences and mathematics. The vocabulary is often unfamiliar and the reasoning is often opaque.

Accordingly, this essay is not easy to follow but will reward the persistent reader. Weyl identifies four issues that we need to confront. (1) What is the mechanism of existence? (2) What is the foundation of quantum mechanics? (3) What is the nature of the continuum of real numbers? (Weyl was a supporter of Brouwer's intuitionism.) (4) What is the nature of time? Of these he regards human existence as the mystery of mysteries. Weyl, incidentally, tacitly rejects positivism as the ultimate answer to reality or the realm of being.

Many in the scientific arena will cling tenaciously to scientific knowledge and eschew philosophic theories, but a moment's reflection on the history of scientific ideas will point to the fact that as science penetrates more deeply into the mysteries of knowledge, many questions are raised. The great discoveries of Newton have been amplified by the deep insights of relativity theory; the simple theories governing the nature of matter and radiation have resulted in the imponderables of quantum theory. Even the apparently straightforward concepts of geometry and logic have given way to more subtle and mystifying insights. Moreover, knowledge as in the case of quantum theory, can be elusive since it is *perturbed* by those seeking it. Using a metaphor from John Bunyan, Weyl asks how we can extricate ourselves from this "Slough of Despond".

Finally Weyl asks: Is there a common essence that threads its way through the many different disciplines? Or are these studies destined to go their separate ways with little to bind them together? The past herculean efforts to bring an essential order to this question remain unfulfilled.

The essay is the text of a lecture given by Weyl on the occasion of the bicentennial of Columbia University in New York, 1954.

The present solemn occasion on which I am given the honor to address you on our general theme "The Unity of Knowledge" reminds me, you will presently see why, of another Bicentennial Conference, held 14 years ago by our neighborly university in the city of Brotherly Love. The words with which I started there a talk on "The Mathematical Way of Thinking" sound like an anticipation of today's topic; I repeat them: "By the mental process of thinking we try to ascertain truth; it is our mind's effort to bring about its own enlightenment by evidence." Hence, just as truth itself and the experience of evidence, it is something fairly uniform and universal in character. Appealing to the light in our innermost self, it is neither reducible to a set of mechanically applicable rules, nor is it divided into water-tight compartments like historical, philosophical, mathematical thinking, etc. True, nearer the surface there are certain techniques and differences; for instance, the procedures of fact-finding in a courtroom and in a physical laboratory are conspicuously different." The same conviction was more forcibly expressed by the father of our Western philosophy, Descartes, who said: "The sciences taken all together are identical with human wisdom, which always remains one and the same, however applied to different subjects, and suffers no more differentiation proceeding from them than the light of the Sun experiences from the variety of the things it illumines."

But it is easier to state this thesis in general terms than to defend it in detail when one begins to survey the various branches of human knowledge. Ernst Cassirer, whose last years were so intimately connected with this university, set out to dig for the root of unity in man by a method of his own, first developed in his great work *Philosophie der symbolischen Formen.* The lucid *Essay on Man* written much later in this country and published by the Yale University Press in 1944, is a revised and condensed version. In it he tries to answer the question "What is man?" by a penetrating analysis of man's cultural activities and creations: language, myth, religion, art, history, science. As a common feature of all of them he finds: the symbol, symbolic representation. He sees in them "the threads which weave the symbolic net, the tangled net of human experience." "Man," he says, "no longer lives in a merely physical universe, he lives in a symbolic universe." Since "reason is a very inadequate term with which to comprehend the forms of man's cultural life in all their richness and variety," the definition of man as the *animal rationale* had better be replaced by defining him as an *animal symbolicum*. Investigation of these symbolic forms on the basis of appropriate structural categories should ultimately tend towards displaying them as "an organic whole tied together not by a *vinculum substantiale*, but a *vinculum functionale*." Cassirer invites us to look upon them ''as so many variations on a common theme," and sets as the philosopher's task "to make this theme audible and understandable." Yet much as I admire Cassirer's analyses, which betray a mind of rare universality culture and intellectual experience, their sequence, as one follows them in his book, resembles more a suite of bourrées, sarabands, menuets and gigues than variations on a single theme. In the concluding paragraph he himself emphasizes "the tensions and frictions, the strong contrasts and deep conflicts between the various powers of man, that cannot be reduced to a common denominator." He then finds consolation in the thought that "this multiplicity and disparateness does not denote discord or disharmony," and his last word is that of Heraclitus: "Harmony in contrariety, as in the case of the bow and the lyre." Maybe, man cannot hope to be more than that; but am I wrong when I feel that Cassirer quits with a promise unfulfilled?

In this dilemma let me now first take cover behind the shield of that special knowledge in which I have experience through my own research: the natural sciences including mathematics. Even here doubts about their methodical unity have been raised. This, however, seems unjustified to me. Following Galileo, one may describe the method of science in general terms as a combination of passive observation refined by active experiment with that symbolic construction to which theories ultimately reduce. Physics is the paragon. Hans Driesch and the holistic school have claimed for biology a methodical approach different from, and transcending, that of physics. However, nobody doubts that the laws of physics hold for the body of an animal or myself as well as for a stone. Driesch's attempts to prove that the organic processes are incapable of mechanical explanation rest on a much too narrow notion of mechanical or physical explanation of nature. Here quantum physics has opened up new possibilities. On the other side, wholeness is not a feature limited to the organic world. Every atom is already a whole of quite definite structure; its organization is the foundation of possible organizations and structures of the utmost complexity. I do not suggest that we are safe against surprises in the future development of science. Not so long ago we had a pretty startling one in the transition from classical to quantum physics. Similar future breaks may greatly affect the. epistemological interpretation, as this one did with the notion of causality; but there are no signs that the basic method itself, symbolic construction combined with experience, will change.

It is to be admitted that on the way to their goal of symbolic construction scientific theories pass preliminary stages, in particular the classifying or morphological stage. Linnaeus' classification of plants, Cuvier's comparative anatomy are early examples; comparative linguistics or jurisprudence are analogues in the historical sciences. The features which natural science determines by experiments, repeatable at any place and any time, are *universal;* they have that empirical necessity which is possessed by the laws of nature. But beside this domain of the necessary there remains a domain of the *contingent.* The one cosmos of stars and diffuse matter, Sun and Earth, the plants and animals living on earth, are accidental or singular phenomena. We are interested in their evolution. Primitive thinking even puts the question "How did it come about" before the question "How is it." All history in the proper sense is concerned with the development of one singular phenomenon: human civilization on earth. Yet if the experience of natural science accumulated in her own history has taught one thing, it is this, that in its field knowledge of the laws and of the inner constitution of things must be far advanced before one may hope to understand or hypothetically reconstruct their genesis. For want of such knowledge as is now slowly gathered by genetics, the speculations on pedigrees and phylogeny let loose by Darwinism in the last decades of the 19th century were mostly premature. Kant and Laplace had the firm basis of Newton's gravitational law when they advanced their hypotheses about the origin of the planetary system.

After this brief glance at the methods of natural science, which are the same in all its branches, it is time now to point out the limits of science. The riddle posed by the double nature of the ego certainly lies beyond those limits. On the one hand, I am a real individual man; born by a mother and destined to die, carrying out real physical and psychical acts, one among many (far too many, I may think, if boarding a subway during rush hours). On the other hand, I am "vision" open to reason, a self-penetrating light, immanent sense-giving consciousness, or however you may call it, and as such unique. Therefore I can say to myself both: "I think, I am real and conditioned" as well as "I think, and in my thinking I am free." More clearly than in the acts of volition the decisive point in the problem of freedom comes out, as Descartes remarked, in the theoretical acts. Take for instance the statement $2 + 2 = 4$: not by blind natural causality, but because I *see* that $2 + 2 = 4$ does this judgement as a real psychic act form itself in me, and do my lips form these words: two and two make four. Reality or the realm of Being is not closed, but open toward Meaning in the ego, where Meaning and Being are merged in indissoluble union–though science will never tell us how. We do not see through the real origin of freedom.

And yet, nothing is more familiar and disclosed to me than this mysterious "marriage of light and darkness," of self-transparent consciousness and real being that I am myself. The access is my knowledge of myself from within, by which I am aware of my own acts of perception, thought, volition, feeling and doing, in a manner entirely different from the theoretical knowledge that represents the "parallel" cerebral processes in symbols. This inner awareness of myself is the basis for the more or less intimate understanding of my fellow men, whom I acknowledge as beings of my own kind. Granted that I do not know of their consciousness in the same manner as of my own, nevertheless my "interpretative" understanding of it is apprehension of indisputable adequacy. As hermeneutic interpretation it is as characteristic for the historical, as symbolic construction is for the natural sciences. Its illumining light not only falls on my fellow-men; it also reaches, though with ever increasing dimness and incertitude, deep into the animal kingdom. Kant's narrow opinion that we can feel compassion, but cannot share joy with other living creatures, is justly ridiculed by Albert Schweitzer who asks: "Did he never see a thirsty ox coming home from the fields, drink?" It is idle to disparage this hold on nature "from within" as anthropomorphic and elevate the objectivity of theoretical construction, though one must admit that understanding, for the very reason that it is *concrete* and *full,* lacks the freedom of the "hollow symbol." Both roads run, as it were, in opposite directions: what is darkest for theory, man, is the most luminous for the understanding from within; and to the elementary inorganic processes, that are most easily approachable by theory, interpretation finds no access whatever. In biology the latter may serve as a guide to important problems, although it will not provide an objective theory as their solution. Such teleological statements as "The hand is there to grasp, the eye to see" drive us to find out what internal material organization enables hand and eye, according to the physical laws (that hold for them as for any inanimate object), to perform these tasks.

I will not succumb to the temptation of foisting Professor Bohr's idea of complementarity upon the two opposite modes of approach we are discussing here. However, before progressing further, I feel the need to say a little more about the constructive procedures of mathematics and physics.

Democritus, realizing that the sensuous qualities are but effects of external agents on our sense organs and hence mere apparitions, said: "Sweet and bitter, cold and warm, as well as the colors, all these exist but in opinion and not in reality; what really exist are unchangeable particles, atoms, which move in empty space." Following his lead, the founders of modern science, Kepler, Galileo, Newton, Huygens, with the approval of the philosophers, Descartes, Hobbes, Locke, discarded the sense qualities, on account of their subjectivity, as building material of the objective world which our perceptions reflect. But they clung to the objectivity of space, time, matter, and hence of motion and the corresponding geometric and kinematic concepts. Thus Huygens, for instance, who developed the undulatory theory of light, can say with the best of conscience that colored light beams are *in reality* oscillations of an ether consisting of tiny particles. But soon the objectivity of space and time also became suspect. Today we find it hard to realize why their intuition was thought particularly trustworthy. Fortunately Descartes' analytic geometry had provided the tool to get rid of them and to replace them by numbers, i.e., mere symbols. At the same time one learned how to introduce such concealed characters, as, e.g., the inertial mass of a body, not by defining them explicitly, but by postulating certain simple laws to which one subjects the observation of reacting bodies. The upshot of it all is a purely symbolic construction that uses as its material nothing but mind's free creations: symbols. The monochromatic beam of light, which for Huygens was in reality an ether wave, has now become a formula expressing a certain undefined symbol F, called electromagnetic field, as a mathematically defined function of four other symbols x, y, z, t, called space-time coordinates. It is evident that now the words "in reality" must be put between quotation marks; who could seriously pretend that the symbolic construct is the true real world? Objective Being, reality, becomes elusive; and science no longer claims to erect a sublime, truly objective world above the Slough of Despond in which our daily life moves. Of course, in some way one must establish the connection between the symbols and our perceptions. Here, on the one hand, the symbolically expressed laws of nature (rather than any explicit "intuitive" definitions of the significance of the symbols) play a fundamental role, on the other hand the concretely described procedures of observation and measurement.

In this manner a theory of nature emerges which only as a whole can be confronted with experience, while the individual laws of which it consists, when taken in isolation, have no verifiable content. This discords with the traditional idea of truth, which looks at the relation between Being and Knowing from the side of Being, and may perhaps be formulated as follows: "A statement points to a fact, and it is true if the fact to which it points is so as it states." The truth of physical theory is of a different brand.

Quantum theory has gone even a step further. It has shown that observation always amounts to an uncontrollable intervention, since measurement of one quantity irretrievably destroys the possibility of measuring certain other quantities. Thereby the objective Being which we hoped to construct as one big piece of cloth each time tears off; what is left in our hands are—rags.

The notorious man-in-the-street with his common sense will undoubtedly feel a little dizzy when he sees what thus becomes of that reality which seems to surround him in such firm, reliable and unquestionable shape in his daily life. But we must point out to him that the constructions of physics are only a natural prolongation of operations his own mind performs (though mainly unconsciously) in perception, when e. g. the solid shape of a body constitutes itself as the common source of its various perspective views. These views are conceived as appearances, for a subject with its continuum of possible positions, of an entity on the next higher level of objectivity: the three-dimensional body. Carry on this "constitutive" process in which one rises from level to level, and one will land at the symbolic constructs of physics. Moreover, the whole edifice rests on a foundation which makes it binding for all reasonable thinking: of our complete experience it uses only that which is unmistakably *aufweisbar.*

Excuse me for using here the German word. I explain it by reference to the foundations of mathematics. We have come to realize that isolated statements of classical mathematics in most cases make as little sense as do the statements of physics. Thus it has become necessary to change mathematics from a system of meaningful propositions into a game of formulas which is played according to certain rules. The formulas are composed of certain clearly distinguishable symbols, as concrete as the men on a chess board. Intuitive reasoning is required and used merely for establishing the consistency of the game—a task which so far has only partially been accomplished and which we may never succeed in finishing. The visible tokens employed as symbols must be, to repeat Hilbert's words, "recognizable with certainty, independently of time and place, and independently of minor differences and the material conditions of their execution (e.g., whether written by pencil on paper or by chalk on blackboard)." It is also essential that they should be reproducible where—and whenever needed. Now here is the prototype of what we consider as *aufweisbar*, as something to which we can point *in concreto*. The inexactitude which is inseparable from continuity and thus clings inevitably to any spatial configurations is overcome here in principle, since only clearly distinguishable marks are used and slight modifications are ignored "as not affecting their identity." (Of course, even so errors are not excluded.) When putting such symbols one behind the other in a formula, like letters in a printed word, one obviously employs space and spatial intuition in a way quite different from a procedure that makes space in the sense of Euclidean geometry with its exact straight lines etc. one of the bases on which knowledge rests, as Kant does. The *Aufweisbare* we start with is not such a pure distillate, it is much more concrete.

Also the physicist's measurements, e.g., reading of a pointer, are operations performed in the *Aufweisbaren*—although here one has to take the approximate character of all measurements into account. Physical theory sets the mathematical formulas consisting of symbols into relation with the results of concrete measurements.

At this juncture I wish to mention a collection of essays by the mathematician and philosopher Kurt Reidemeister published by Springer in 1953 and 1954 under the titles *Geist und Wirklichkeit* and *Die Unsachlichkeit der Existenzphilosophie*. The most important is the essay "Prolegomena einer kritischen Philosophie" in the first volume. Reidemeister is positivist in as much as he maintains the irremissible nature of the factual which science determines; he ridicules (rightly, I think) such profound sounding but hollow evocations as Heidegger indulges in, especially in his last publications. On the other hand, by his insistence that science does not make use of our full experience, but selects from it that which is *aufweisbar,* Reidemeister makes room for such other types of experience as are claimed by the windbags of profundity as their proper territory: the experience of the indisposable significant in contrast to the disposable factual. Here belongs the intuition through and in which the beautiful, whether incorporated in a vase, a piece of music or a poem, appears and becomes transparent, and the reasonable experience governing our dealings and communications with other people; an instance: the ease with which we recognize and answer a smile. Of course, the physical and the aesthetic properties of a sculpture are related to each other; it is not in vain that the sculptor is so exacting with respect to the geometric properties of his work, because the desired aesthetic effect depends on them. The same connection is perhaps even more obvious in the acoustic field. Reidemeister, however, urges us to admit our *Nicht-Wissen,* our not knowing how to combine these two sides by theory into one unified realm of Being—just as we cannot see through the union of I, the conditioned individual, and I who thinking am free. This *Nicht-Wissen* is the protecting wall behind which he wants to save the indisposable significant from the grasp of hollow profundity and restore our inner freedom for a genuine apprehension of ideas. Maybe, I overrate Reidemeister's attempt, which no doubt is still in a pretty sketchy state, when I say that, just as Kant's philosophy was based on, and made to fit, Newton's physics, so his attempt takes the present status of the foundations of mathematics as its lead. And as Kant supplements his Critique of Pure Reason by one of practical reason and of aesthetic judgement, so leaves Reidemeister's analysis room for other experiences than science makes use of, in particular for the hermeneutic understanding and interpretation on which history is based.

Let me for the few remarks I still want to make adopt the brief terms science and history for natural and historical sciences (*Natur und*

Geistes Wissenschaften). The first philosopher who fully realized the significance of hermeneutics as the basic method of history was Wilhelm Dilthey. He traced it back to the exegesis of the Holy Script. The chapter on history in Cassirer's "Essay on man" is one of the most successful. He rejects the assumption of a special historical logic or reason as advanced by Windelband or more recently and much more impetuously by Ortega y Gasset. According to him the essential difference between history and such branches of science as e. g. palaeontology dealing with singular phenomena lies in the necessity for the historian to interpret his "petrefacts," his monuments and documents, as having a symbolic content.

Summarizing our discussion I come to this conclusion. At the basis of all knowledge there lies: (1) *Intuition,* mind's originary act of "seeing" what is given to him; limited in science to the *Aufweisbare,* but in fact extending far beyond these boundaries. How far one should go in including here the *Wesensschau* of Husserl's phenomenology, I prefer to leave in the dark. (2) *Understanding and expression.* Even in Hilbert's formalized mathematics I must understand the directions given me by communication in words for how to handle the symbols and formulas. Expression is the active counterpart of passive understanding. (3) *Thinking the possible.* In science a very stringent form of it is exercised when, by thinking out the possibilities of the mathematical game, we try to make sure that the game will never lead to a contradiction; a much freer form is the imagination by which theories are conceived. Here, of course, lies a source of subjectivity for the direction in which science develops. As Einstein once admitted, there is no logical way leading from experience to theory, and yet the decision as to which theories are adopted turns out ultimately to be unambiguous. Imagination of the possible is of equal importance for the historian who tries to re-enliven the past. (4) On the basis of intuition, understanding and thinking of the possible, we have in science: certain practical actions, namely the *construction* of symbols and formulas on the mathematical side, the construction of the measuring devices on the empirical side. There is no analogue for this in history. Here its place is taken by *hermeneutic interpretation,* which ultimately springs from the inner awareness and knowledge of myself. Therefore the work of a great historian depends on the richness and depth of his own inner experience. Cassirer finds wonderful words for Ranke's intellectual and imaginative, not emotional, sympathy, the universality of which enabled him to write the history of the Popes and of the Reformation, of the Ottomans and the Spanish Monarchy.

Being and Knowing, where should we look for unity? I tried to make clear that the shield of Being is broken beyond repair. We need not shed too many tears about it. Even the world of our daily life is not *one,* to the extent people are inclined to assume; it would not be difficult to show up some of its cracks. Only on the side of Knowing there may be unity. Indeed, mind in the fullness of its experience has unity. Who says "I" points to it. But just because it is unity, I am unable to represent it otherwise than by such characteristic actions of the mind mutually supporting each other as I just finished enumerating. Here, I feel, I am closer to the unity of the luminous center than where Cassirer hoped to catch it: in the complex symbolic creations which this lumen built up in the history of mankind. For these, and in particular myth, religion, and alas!, also philosophy, are rather turbid filters for the light of truth, by virtue, or should I say, by vice of man's infinite capacity for self-deception.

What else than turbidity could you then have expected from a philosophical talk like this? If you found it particularly aimless, please let me make a confession before asking for your pardon. The reading of Reidemeister's essays has caused me to think over the old epistemological problems with which my own writings had dealt in the past; and I have not yet won through to a new clarity. Indecision of mind does not make for coherence in speaking one's mind. But then, would one not cease to be a philosopher, if one ceased to live in a state of wonder and mental suspense?

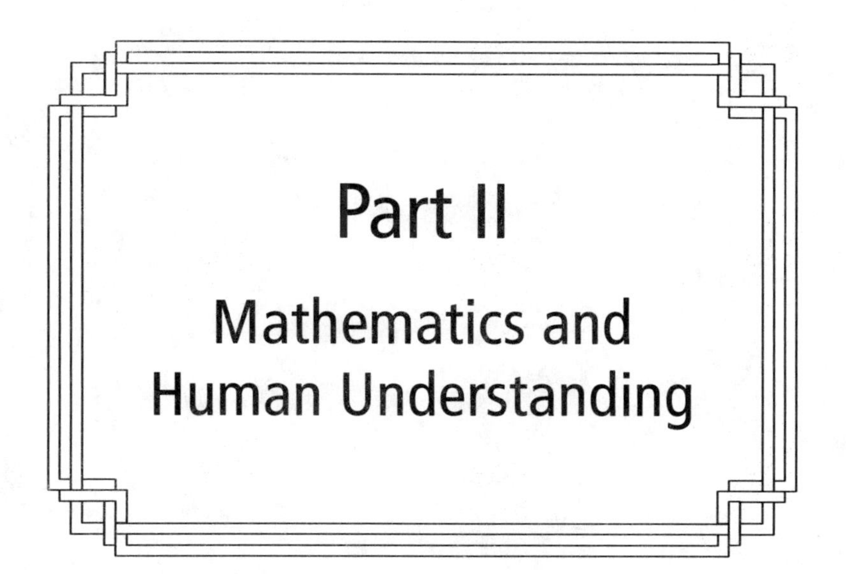

Part II

Mathematics and Human Understanding

"Do not worry about your difficulties in mathematics,
I assure you mine are greater."
—Albert Einstein (1879–1955)

Mathematics and the Arts

Marston Morse

Harold Calvin Marston Morse (1892–1977) was born in Waterville, Maine and it is said that in later life he retained many of the characteristics associated with New Englanders. His parents were Howard Calvin Morse and Ella Phoebe Marston.

From his mother he inherited not only his name but a notable talent for music and from his father habits of thrift and industry. He loved music and was an accomplished pianist and organist, occasionally venturing into the realm of musical composition.

In Waterville he had his formal education from elementary school through Colby College where he was active in sports, becoming an accomplished tennis player.

He received a PhD from Harvard under the supervision of G.D. Birkhoff who readily recognized Morse's mathematical gifts. Although he was offered a traveling fellowship in 1917, he chose instead to serve in the U.S. Army in France. For his valor, the French government awarded him the Croix de Guerre.

He served on the faculties of Cornell and Brown after which he was appointed to the faculty at Harvard. There he remained until his appointment to the newly established Institute for Advanced Study in 1935. He retired in 1962. As with World War I, the Second World War brought Morse once again into service, this time as a civilian consultant in the Office of the Chief of Ordnance. After the war, among other things, he was instrumental in helping to establish the National Science Foundation, an organization that proved to be crucial in the development of American science..

His mathematical achievements were widely recognized and what has come to be called "Morse Theory" is extolled as one of the great achievements of American mathematics up to that time. He served as president of the American Mathematical Society and the International Mathematical Union. He received numerous honors including many honorary degrees both in the U.S. and abroad.

Although he had no formal teaching duties at the Institute, he continued to influence young mathematicians who report enthusiastically on their experiences.

In 1940, he married Louise Jeffreys who bore five children. From an earlier marriage there were two children. She was widely lauded by Morse's friends and associates for her generosity, hospitality, and charm.

Morse manifested a profound solicitude for human welfare, and he supported a number of humanitarian causes. Although one of the characteristics of New Englanders is said to be thrift, in the case of Morse this was accompanied by financial and personal generosity.

Morse had interests in religion, science and the arts. He believed in the basic unity of the creative spirit in all areas of endeavor and this article is Morse's expression of this unity.

Editor's Preface

This charming article by a renowned mathematician deals with a recurrent theme in the evolution of mathematical thought, viz., mathematics and the arts. There are two aspects to this theme. The first is the application of mathematical ideas to the arts—either tacit or explicit. This aspect covers a wide variety of ways in which mathematics injects itself into the arts. In the graphic arts, for example, there is the mathematics of perspective, which had its mathematical realization in projective geometry. Then there is the golden ratio which appears in numerous ways, and to which renaissance artists paid special attention. Some specialists claim that the Greeks had already used this ratio in architecture, such as the Parthenon. There are other instances of applications. In music, we have the construction of mathematical scales dating to Pythagoras, and in relatively recent times there is the Fourier analysis of tones. There are even claims that Mozart used the golden mean in one of his compositions and that Beethoven used it in his fifth symphony! In more modern times, we have the application of permutation groups in the composition of atonal music.

These relations, though interesting to many, are not what Morse has in mind, however. He is suggesting that the same creative and aesthetic forces at work in the arts, are also at work in mathematical invention. This is not a new idea, as a few quotations will testify, but Morse has brought some new insights. To cite some examples from the past, we begin with Aristotle who wrote "the philosophers who claim that the mathematical sciences make no room for either the beautiful or the good are surely mistaken... Beauty on the contrary is the principal object of reasoning in the sciences and in their proofs. The highest forms of beauty are order, symmetry and veracity and it is in mathematics, above all, that these appear." In more recent times, G. Cantor, in an effort to add understanding, proclaimed that the essence of mathematics is freedom. Leonardo asserted that there is no difference between art and science. It is generally recognized that R. Bacon was the first to articulate the difference. Weierstrass said that a true mathematician is a poet. In a somewhat different vein, the German renaissance painter Dürer, one of whose paintings Morse "quotes," wrote as follows: "without Geometry, no one can be or become an absolute artist; the blame is upon their masters who are ignorant of this Art." It is easy to go on in this vein.

An understanding of this creative process is elusive and certainly not at all well understood. Nevertheless, a case can be made that every work of art has a message; indeed that that is the main object of a work of the creator —to convey through a given medium a deep human insight. This message is discerned through an emotional perception opening the paths to a more conscious articulation of this perception. The analysis of this response is essentially what comprises aesthetics. Much investigation has been brought to bear on this subtle process and it is an astonishing fact that there is surprising agreement among commentators on the significance of a given work of art. The aesthetician can endeavor to recount why a piece of art or a piece of writing is meritorious. This effort is often successful in the opinion of many readers. The reality however, is that the awareness of a work of art is far more subtle than can be communicated in words.

As to mathematics, it is interesting that the traditional language of aesthetics is not always appropriate to describing mathematical inventions. Mathematics seems natural to humans and it has been surmised that mathematics can be discovered by pure thought, since in a vague sense it is an outward expression of our inner minds, modified by our sensory experience of the real world. Where do the structures expressed mathematically arise? This is an elusive question but the noted child psychologist Piaget has thrown some light on the subject in his studies of the evolution of mathematical thought in children. How does a mathematician explain, for example, why the prime number theorem is a discovery of great beauty? An association with mathematicians will soon reveal that they use a modified vocabulary of aesthetics to describe mathematical discoveries and inventions. It is relevant to say that mathematics to a mathematician is a source of emotion which arouses deep seated subterranean responses. These affective reactions to some mathematical results are not dissimilar to the response on viewing the Pietà or listening to the Choral Symphony but have, nevertheless, relevant differences.

In recent times neurologists have used PET scans to show vivid pictures of brain patterns when the subject is engaged in various activities. Morse would probably agree that a mathematician engaged in doing mathematics exhibits patterns similar to those of artists engaged in the process of their creative art.

In summary, drawing on his personal experiences, Morse has added some notably worthwhile insights into all these matters.

The article is reprinted from the *Bulletin of the Atomic Scientists*, Vol. X, No. 2, Feb. 1959. It was presented at a conference on Robert Frost held at Kenyon College in 1959.

To talk about art other than in the impersonal sense of history, is to talk about the moments when one has been confronted with beauty. Every essay on art that lights a hidden niche has its source in the life of the writer. You will then perhaps understand why I start with the mood of my childhood.

One hundred miles northeast of Derry, New Hampshire, lie the Belgrade Lakes, and out of the last and longest of these lakes flows the Messalonskee. I was born in its valley, "north of Boston" in the land of Robert Frost. The "Thawing Wind" was there, the "Snow," the "Birches," and the "Wall" that had to be mended: I was born on a sprawling farm cut by a pattern of brooks that went nowhere—and then somewhere. A hundred acres of triangles of timothy and clover, and twisted quadrilaterals of golden wire grass, good to look at, and good riddance. At ten I combed it all with horse and rake, while watching the traffic of mice beneath the horse's feet.

All that Frost has described was there—the meanness and generosity of men and women. A neighbor's house burned down in a wind, and everyone knew who held the grudge. The woman who must have killed her lover (so everyone thought) stood up in prayer meeting and testified, and there was no more judgment against her than was proper. The autumn winds were the prelude to the loon's strange song. There was time to think in the winter, to like some things better than others.

My mother's world was the world of music, and her world became mine. At thirteen I was playing the organ in church and wished the time in summer to study and practice. Somewhat reluctantly my father conceded me the mornings. He said that the grass was too wet to rake in the morning, and that I could walk the three miles from church to farm. And so I learned some of the Bach Fugues for the organ, and the moving Sonatas which Mendelssohn had written to honor the memory of Bach.

Grecian art first became real to me in the shop of an old cabinet maker. I began to learn from him about cabinet making, and the history of his art. Sheraton chairs and tables were scattered about his shop, with their fluted columns and acanthus leaves. It came to me, all of a sudden, that these were fragments of Grecian temples.

Mathematics the Sister of the Arts

There was a copy of the *Cabinet Maker of Sheraton* in a remote library. It started with descriptive geometry and continued with a theory of ornaments. Cornices were constructed with ruler and compass; symmetry and perfection reigned throughout. Here was a meeting of mathematics and art, something final and universal, as it seemed to me then. It was very alive, because it was so new. But it was not mathematics as I know it today, and as it should be known; it was matter without the spirit. I made the same mistake that artists have made since the time of the Greeks, and placed mathematics alongside of the arts as their *handmaiden.* It is a humble and honorable position and very necessary; for one must begin with exactness in all the arts. But mathematics is the *sister,* as well as the *servant* of the arts and is touched with the same madness and genius. This must be known.

There was a German painter and engraver born in the fifteenth century with the name of Albrecht Dürer who wanted mathematics to be more than a handmaiden of art. His discontent on this account was unique among artists of all time. More completely than any other artist he formulated the rules of symmetry, perspective, and proportion, and used them in his art. But any one who thinks Dürer's spirit is bound by rules is mistaken. There is almost a shock in passing from his rugged, first engravings to the radiant classical beauty and slender proportions of his *Adam and Ev*e of 1507.

Dürer was a creative mathematician as well as an artist. He wanted his geometric theories to measure up to his art. His great engraving *Melencolia I* is a psychological self-portrait. The perplexed and thoughtful heroine is the figure of geometry. Everything I have to say today is hidden in this engraving or may be derived from it by projection into the future. Let me quote from my colleague Erwin Panofsky. "The engraving *Melencolia I*", he says, "... typifies the artist of the Renaissance who respects practical skill, but longs all the more fervently for mathematical theory—who feels 'inspired' by celestial influences and eternal ideas, but suffers all the more deeply from his human frailty and intellectual finiteness ..." Dürer was an artist-geometer, and one who suffered from the very limitations of the discipline he loved. In his younger days, when he prepared the engraving *Adam and Eve*, he had hoped to capture absolute beauty by means of a ruler and a compass. Shortly before he composed the *Melencolia* he was forced to admit: "But what absolute beauty is, I know not. Nobody knows it except God."

In his dependence upon geometry Dürer was inspired by Leonardo but repudiated by Michelangelo. Later artists followed Dürer only half way or not at all; it is indeed hard to follow an inspiration. Leonardo himself had little of Dürer's divine discontent.

Back of Dürer and Leonardo in the distant past stands the Roman architect and geometer Vitruvius. The Mesopotamian artists also looked on geometry as an aid to art, and this was well known to the prophet Isaiah. Chapter 44 of Isaiah is written against idolatry; it is also an essay on aesthetics. The thirteenth verse reads: "The carpenter stretcheth out his rule; he marketh it out with a line; he fitteth it with planes, and he marketh it out with the compass, and maketh it after the figure of a man, according to the beauty of a man; that it may remain in the house."

Isaiah would minimize geometry in the arts, Dürer would maximize it. Neither Isaiah nor Dürer was content.

Let us turn to the relation between mathematics and music. The evolution of the scales, from the archaic sequences of tones of Euripides to

Dürer's *Melencolia I*

the whole tone scale of Debussy, shows that mathematics and music have much in common. And there is also the arithmetical basis for harmony. It is not too difficult to compose in the technical scheme of Debussy and thereby to get some of his naturalistic effects, but no one can explain the profound difference between the opera, *Pelléas et Mélisande,* on the one hand, and *Tristan und Isolde,* on the other, by reference to whole tone scales or any other part of musical theory.

Geometric form imposed on music can have a null effect. As an example, I shall compare the First Prelude of Bach, as found in the *Well-Tempered Clavichord,* with the First Prelude of Chopin. The First Prelude of Bach is without melody, and consists of repeating ascending arpeggios with similar form and length. It is intended that the effect shall be harp-like. The musical text as a whole exhibits a design that appears in no one of the other forty-eight preludes. Looked at geometrically, the First Prelude of Chopin has a very similar geometric design, and if the Chopin prelude is played an octave higher than written, with perfect evenness of tone and tempo, the actual musical similarity of the two preludes is most striking. If, however, the Chopin prelude is played with the color and pulsating rhythm which it demands, all similarity to the Bach disappears.

Most convincing to me of the spiritual relations between mathematics and music, is my own very personal experience. Composing a little in an amateurish way, I get exactly the same elevation from a prelude that has come to me at the piano, as I do from a new idea that has come to me in mathematics.

Nature of Affinity

My thesis is prepared. It is that *the basic affinity between mathematics and the arts is psychological and spiritual and not metrical or geometrical.*

The first essential bond between mathematics and the arts is found in the fact that discovery in mathematics is not a matter of logic. It is rather the result of mysterious powers which no one understands, and in which the unconscious recognition of beauty must play an important part. Out of an infinity of designs a mathematician chooses one pattern for beauty's sake, and pulls it down to earth, no one knows how. Afterwards the logic of words and of forms sets the pattern right. Only then can one tell someone else. The first pattern remains in the shadows of the mind.

All this is like Robert Frost's "figure a poem makes." The poet writes: "I tell how there may be a better wildness of logic, than of inconsequence. But the logic is backward, in retrospect after the act. It must be more felt, than seen ahead like prophecy." Or again, "For me the initial delight is in the surprise of remembering something I didn't know I knew. I am in a place, in a situation, as if I had materialized from cloud, or risen out of the ground."

Compare this with the account of how the French mathematician Henri Poincaré came to make one of his greatest discoveries. While on a geologic excursion a mathematical idea came to him. As he says it came "without anything in my former thoughts seeming to have paved the way." He did not then have the time to follow up this idea. On returning from his geologic excursion he sought to verify the idea. He had no immediate success, and turned to certain other questions which interested him, and which seemed at the time to have no connection with the idea which he wished to verify. Here again he was unsuccessful. Disgusted with his failure he spent a few days at the seaside and thought of something else. One morning while walking on the bluff the final solution came to him with the same characteristics of brevity and suddenness as he had experienced on sensing the initial idea, and quite remarkably he had a sense of complete certainty. He made his great discovery. [Poincaré's account may be read in his essay appearing below].

An account of Gauss is similar. He tells how he came to establish a theorem which had baffled him for two years. Gauss writes: "Finally, two days ago, I succeeded, not on account of my painful efforts, but by the grace of God. Like a sudden flash of lightning, the riddle happened to be solved. I myself cannot say, what was the conducting thread, which connected what I previously knew, with what made my success possible."

These words of Frost, Poincaré, and Gauss show how much artists are in agreement as to the psychology of creation.

A second affinity between mathematicians and other artists lies in a psychological necessity under which both labor. Artists are distinguished from their fellows who are not artists by their overriding instinct of self-preservation as creators of art. This is not an economic urge as everyone knows who has a variety of artist friends. I shall illustrate this by the case of Johann Sebastian Bach and his son Philipp Emanuel.

Johann Sebastian's work culminates and closes a religious and musical epoch. It is inconceivable that Philipp Emanuel could have continued as a composer in the same sense as his father and have lived as an artist. He did in fact reject his father's musical canons. There is considerable evidence that his environment called for a new musical spirit. History justifies Philipp Emanuel; Mozart said of him, "He is the father, we are the children"; Haydn was inspired by him and Beethoven admired him. With all this to his credit, posterity can perhaps forgive him for calling his father an old whig.

Quite analogous to the son's turning away from his father is the story of the relation of mathematician Henri Poincaré to his younger colleague, Lebesgue. Poincaré had used the materials of the nineteenth-century mathematics to revolutionize much of mathematics. He had gone so far in mathematics that it is doubtful whether his younger colleagues in France could go on in the same sense without introducing essentially new techniques. This was in fact what several of them did. One of the new fields was what is called "set theory," and one of the innovators Lebesgue.

Poincaré criticized the members of the new school rather severely. It is on record that at a Congress in Rome he made this prediction. "Later generations will regard set theory as a malady from which one has recovered." (One may remark parenthetically that the history of art records many maladies from which art has recovered.)

The response of Lebesgue to Poincaré was given on his elevation to a Professorship at the Collège de France. An older eminent colleague had praised the school of Lebesgue. Lebesgue made public reference to the "precious encouragement which had largely compensated for the reproaches" which his school had had to suffer.

I regard the reactions of both Poincaré and Lebesgue as dictated by instincts of self-preservation, typical of the artist. Such self-preservation was clearly to the advantage of mathematics as well. I am also one of the few mathematicians who think that Poincaré as well as Lebesgue was right, in that mathematics will return more completely to the great ideas of Poincaré with full appreciation of the innovations of Lebesgue, but with a truer understanding of the relation of mathematical technique to mathematical art.

Before coming to the third type of evidence of the affinity of mathematics with the rest of the arts it might be well to ask what is it that a mathematician wants as an artist. I believe that he wishes merely to understand and to create. He wishes to understand, simply, if possible—but in any case to understand; and to create, beautifully, if possible—but in any case to create. The urge to understand is the urge to embrace the world as a unit, to be a man of integrity in the Latin meaning of the word. A world which values great works of art, music, poetry, or mathematics, can only approve and honor the urge of any man capable of such activities, to create.

The third type of evidence of the affinity of mathematics with the arts is found in the comparative history of the arts. The history of the arts is the history of recurring cycles and sharp antitheses. These antitheses set pure art against mixed art, restraint against lack of restraint, the transient against the permanent, the abstract against the nonabstract. These antitheses are found in all of the arts, including mathematics.

In particular the antithesis of pure art and mixed art is very much in evidence in the relations between poetry and music. There have been those who wished to keep poetry and music separate at all times. Plato took sides when he said, "Poetry is the Lord of the Lyre," and music had to fight a long battle to obtain complete autonomy.

Quite analogously in mathematics there are those who would like to keep algebra and geometry apart, or would like to subordinate one to the other. The battle became acute when the discovery of analytic geometry by Descartes made it finally possible to represent all geometry by algebra. The battle between algebra and geometry has been waged from antiquity to the present.

Grecian art was of course restrained and a departure from restraint has always brought a reaction. Berlioz gave an example of extreme lack of restraint. To get the maximum effect of Doomsday trumpets in *The Last Day of the World,* Berlioz devised four full-fledged brass bands to play high in the four corners of St. Peters. One American composer even wanted to fire cannon on the beat.

Mathematicians too, are often unrestrained. In this direction are the grandiose cosmologies with more generality than reality. These fantasies are sometimes based neither on nature or logic. Mathematicians of today are perhaps too exuberant in their desire to build new logical foundations for everything. Forever the foundation and never the cathedral. Logic is now so well understood that the laying of foundations is not very difficult. The thing has gone so far that one of my Polish colleagues recently suggested that the right to lay foundations should be rationed, or put on the basis of the right to build one foundation for every genuine classical effort.

The antithesis between logic and intuition manifested itself in the days of the Greeks. Pythagoras had a mystical preference for whole numbers. The irrational numbers were not understood by the Greeks and hence avoided as much as possible. History has made a full turn and the nineteenth century saw the meteoric rise of a more sophisticated Pythagoras by the name of Kronecker. Kronecker laid down the rule "all results of mathematical analysis must ultimately he expressible in properties of integers."

This proclamation cut deeply into the life and work of Kronecker's colleague Weierstrass. Here are a few lines from Weierstrass's reproach.

> "But the worst of it is that Kronecker uses his authority to proclaim, that all those who up to now have labored to establish the theory of functions, are sinners before the Lord—truly it is sad, and it fills me with a bitter grief, to see a man, whose glory is

without flaw, let himself be driven by the well justified feeling of his own worth, to utterances whose injurious effect upon others he seems not to perceive."

The human documents which I have put before you are not concerned with processes which a machine can duplicate. One cannot decide between Kronecker and Weierstrass by a calculation. Were that the case, many of us would turn to another and truer art. As Dürer knew full well, there is a center and final substance in mathematics whose perfect beauty is rational, but rational "in retrospect." The discovery which comes before, those rare moments which elevate man, and the searchings of the heart which come after are not rational. They are gropings filled with wonder and sometimes sorrow.

Often, as I listen to students as they discuss art and science, I am startled to see that the "science" they speak of and the world of science in which I live are different things. The science that they speak of is the science of cold newsprint, the crater-marked logical core, the page that dares not be wrong, the monstrosity of machines, grotesque deifications of men who have dropped God, the small pieces of temples whose plans have been lost and are not desired, bids for power by the bribe of power secretly held and not understood. It is science without its penumbra or its radiance, science after birth, without intimation of immortality.

The creative scientist lives in "the wildness of logic" where reason is the handmaiden and not the master. I shun all monuments that are coldly legible. I prefer the world where the images turn their faces in every direction, like the masks of Picasso. It is the hour before the break of day when science turns in the womb, and waiting, I am sorry that there is between us no sign and no language except by mirrors of necessity. I am grateful for the poets who suspect the twilight zone.

The more I study the interrelations of the arts the more I am convinced that every man is in part an artist. Certainly as an artist he shapes his own life, and moves and touches other lives. I believe that it is only as an artist that man knows reality. *Reality is what he loves, and if his love is lost it is his sorrow.*

References

Erwin Panofsky. *Albrecht Dürer*. Vol. I, pp. 157–71. Princeton University Press.

———. *The Codex Huygens and Leonardo da Vinci's Art Theory*. See page 107. London: The Warburg Institute.

Curt Sachs. *The Commonwealth of Art*. See page 244 for Berlioz. New York: W. W. Norton and Co., Inc.

Jacques Hadamard. *The Psychology of Invention in the Mathematical Field*. Princeton University Press.

E. T. Bell. *Men of Mathematics*. See chapter on Kronecker. New York: Simon and Schuster.

Morse's Epilogue

This essay was read at a conference in honor of Robert Frost in 1950 at Kenyon College. The subject of the Conference was "The Poet and Reality" and there were four speakers besides myself. Among these was L. A. Strong, a distinguished Irish poet and novelist. I single out this poet because it was certain sentences of his which caused me to rewrite the concluding paragraphs of my essay. Strong read the lines of one of his poems with great feeling and beauty of diction. When he had finished, he had occasion to refer to science, and I have no doubt he had in mind the opening of the atomic era. Placing his hand on his forehead he said, "Science is here," and then placing his hand upon his heart, he continued, "And poetry is here." The poet's words and gestures were pregnant with misunderstanding. Each scientist will have his own interpretation of what the speaker meant. I felt moved to rewrite much of what I had written, in the interest of science, and of poetry.

It is with some misgivings that I conceive of an essay originally intended for an audience of poets as now presented to an audience primarily of physicists. My doubts are somewhat lessened by a belief that many physicists, like mathematicians, are guided in the discovery and shaping of their theories by their sense of harmony and of beauty. I know, however, that there are some mathematicians and presumably some physicists, who do not feel any such guidance or at least do not regard the influence of the aesthetic in their groping for scientific law as important.

I do not understand such reluctance to admit the extra-empirical mental processes, particularly when one approaches such problems as those of particle physics, or of a reformulation of quantum mechanics. One cannot believe that the mathematical forms chosen to represent experimental fact are always uniquely determined by the empirical data. I shall illustrate this by referring to the bases of the general relativity theory. It is clear that this theory is much more flexible and more likely than the Newtonian theory, and that it is consistent with important experimental evidence. But admitting this, the form of the general relativity is not thereby uniquely determined. There is in the background of general relativity theory a tacit assumption that the paths of light which it is desired to represent correspond to a solution of the inverse problem of the calculus of variations, a problem which in general is known to have no solution, and which, with even less mathematical generality, has the geodesic solution which Einstein presupposes. The solution which Einstein has accepted may be a very good approximation to the observed physical universe for bounded space-time but at the same time may be infinitely in error when applied to a region of space-time which is unbounded; that is when applied to a cosmology which is not known to correspond to a closed and bounded universe.

There is here a tension between the aesthetically simple, and the mathematically general (sometimes less simple). One has to learn how properly to resolve this tension. But first of all one must admit the uncertainty and seek to understand it. To the extent to which there is a multiplicity of mathematical forms a priori available to express an empirically anchored physical law, to that extent one must call on further experimental evidence, or logic, or aesthetic judgment.

At stake is not only truth, but freedom. Such freedom of choice as exists must be acknowledged and comprehended, or else it is lost. In a milieu in which freedom of hypothesis is well understood the likelihood of intuitive discovery of high order will certainly be increased. The satisfactions of the physicist and the artist may be combined.

November 20, 1958

"Whoever ... demonstrates his prime truth geometrically, should be believed by all the world, for there we are captured." —Albrecht Dürer (1471–1528)

Intuition, Reason and Faith in Science

George David Birkhoff

George David Birkhoff (1884–1944) was born in Overisel, Michigan. He was one of the leading mathematicians of his generation and was the first native-born American mathematician judged to rank with the renowned Europeans. He achieved this distinction with his formulation and proof of the ergodic theorem and proof of the Poincaré fixed-point theorem —conjectures that had challenged the distinguished scholars of Europe.

His antecedents came to Chicago from Holland in 1871 and his father David earned his way through Rush Medical College and eventually practiced medicine in Chicago. Birkhoff attended Lewis Institute, a privately endowed school with high educational aspirations. It was here that his mathematical gifts became apparent. He was just 18 when he published a paper jointly with H. S. Vandiver. It was related to Fermat's last theorem.

He attended Harvard and upon graduation, enrolled as a graduate student at the University of Chicago. It is relevant to note that at the turn of the century, the University of Chicago was considered preeminent in graduate study and research in mathematics. Here he earned a doctorate under E.H.

Moore with a thesis on asymptotics of differential equations. Among his teachers were Maxime Bôcher and W.F. Osgood. He then taught at Princeton, and Wisconsin, and in 1912 he went to Harvard where he remained the rest of his academic life. In 1907, he married Marjorie Grafius who gave her husband every encouragement. A woman with varied talents, she maintained a home noted for its hospitality.

Apart from his fundamental work in dynamical systems, Birkhoff had wide-ranging mathematical interests including graph theory and more exotically, the mathematics of aesthetics. At the International Congress of Mathematicians held in 1928, he gave a lecture on this theme which, appropriately, was held at the famous Palazzo Vecchio, a historic and elegant building in Florence. His book *Aesthetic Measure* had a broad audience and was widely debated in the halls of artists, poets, and musicians.

He is reputed to have been outgoing, if not gregarious. In his political and social views, it is said that he was somewhat detached from the world about him, his primary focus being on the world of mathematics. He played a significant role in the development of mathematics in the United States. He was especially concerned about the generation of mathematicians coming out in the thirties when, as a consequence of political and economic conditions, employment was very problematical.

He maintained close ties with mathematicians in Germany, Italy, and France, though he took great pride in the achievements of American mathematicians. In the genetics of mathematical talent, it is said that inheritance passes from father-in-law to son-in-law. In the case of Birkhoff, inheritance passed normally from father to his son Garrett who achieved renown in his own right.

Among Birkhoff's distinctions are his service as president of the American Mathematical Society as well as president of the American Association for the Advancement of Science. His premature death in 1944 at the age of 60 was a great loss to the world of mathematics, especially American mathematics.

Editor's Preface

In this article, Birkhoff discusses the roles of intuition, faith and reason in the sciences. He begins by taking as axiomatic that an imperative obligation of society is the determination of the place of humanity in the cosmos. To be sure, this is an arduous task but Birkhoff feels that it must be undertaken. In this pursuit, the discoveries of science play a pivotal role; yet he views as equally valid in this quest, the utterances of poets, philosophers, prophets and one is inclined to add, musicians and artists.

Across the millennia, analyses of the nature of knowledge have developed numerous points of view and have led to the creation of various philosophies, cosmologies, etc. These sometimes pitted science against religion and sometimes endeavored to reconcile the two. A notable example is the controversy provoked by Darwinism. Historically, many positions have been advocated. It is nevertheless interesting that many creative scientists have been comfortable with a dualist position. The great Newton, for example, had no difficulty adhering to his religious convictions and at the same time creating wonderful models of the universe. Indeed he saw in the laws of motion, an affirmation of the glory of God. We could add parenthetically, that he even undertook to determine the date of the Second Coming. Another example is Napier who was revered by many scholars on the continent not for his invention of the logarithm, but for his analysis of the Apocalypse, the Book of Revelation.

A prevalent approach among philosophers and others is what might be called the reductionist point of view. In this outlook, the processes of mathematics are reduced to logic, those of physics to those of mathematics, and so on. Birkhoff recognizes six divisions of science, viz., mathematics, physics, chemistry, biology, psychology and sociology. Birkhoff eschews this empiricist point of view; he argues in favor of the use of faith and intuition in scientific discovery. For the mathematician, after all, the consistency of mathematics is an act of faith. Indeed Hermann Weyl once said that mathematics is consistent—hence God exists; we cannot prove it, hence the devil exists!

Birkhoff sees little to be gained from the reductionist viewpoint. To consider examples: the fact that mental depression is caused by the lack of a chemical in the brain is very helpful in the treatment of that disease but it does not throw much light on the phenomenon as a behavioral problem. Again, the fact that the geometry of space is governed by a non-Euclidean metric is not, in itself, revealing; it is the intuition of the physicist, using that fact, that contributes to a more profound understanding of the universe. In these efforts at reduction, much of the essence of the subject may be lost; there is a sort of *gestalt* which, if depleted, robs the ideas of their substance and may lead to a distortion of the underlying intuition.

Birkhoff concedes that his point of view is debatable and tacitly invites discussion. He gives, as an example, the assertions of Faraday about energy, etc.; these were subsequently partially refuted by Maxwell's mathematical formulation of electromagnetism. In this case the reduction of physical phenomena to mathematical ones proved to be germane.

It is an essay from a deep thinker, which merits our cautious attention.

The article is reprinted from an article of the same name which appeared in the journal *Science* in December, 1938, Vol. 88.

It was originally given as an address by Birkhoff, as president of the American Association for the Advancement of Science, in December of 1938, in Richmond, Virginia.

From the earliest times scientific ideas even when crudely conceived have been of immeasurable importance, not only for man's material advancement and control over nature, but also in modifying and expanding his philosophic and religious outlook. In the effort to obtain a better understanding of his place in the cosmos, he is compelled to proceed largely by considerations of analogy based upon supposed or actual fact. And so he turns more and more toward the ever-widening vistas suggested by science in its continual discoveries of new truth.

Today the significance of science as a principal source of revelation is almost universally recognized. Thus recently, on behalf of Pope Pius XI, Cardinal Pacelli spoke before the Pontifical Academy of Sciences concerning the enlightenment that comes from "the potent streams of the natural and rational sciences and the great river of revealed wisdom." He said that the former are found "wherever man looks for and finds truth." As for "the great river of revealed wisdom," is it not to be found in all the absolutely sincere utterances of poets, philosophers and prophets, based on the relevant knowledge of their day and made after deepest meditation? It would seem that such utterances are in essence similar to the pronouncements of the scientist. Is not the vague, prophetic conjecture of Pythagoras that nature is mathematical as true as Newton's more precise law of gravitation? From this point of view, the great streams of revelation seem to merge insensibly into one.

Nevertheless, the immediate effect of scientific advances is often very disquieting. The strong opposition long shown to the Darwinian theory of evolution bears witness to this fact. Similarly at the present day the ever-increasing number of uncoordinated theories and mechanical inventions confuses and chills many of us. Man is felt to be a mere tragic detail in a vast incomprehensible whole, and our old sense of values seems to become less and less real.

To persist in such an attitude of discouragement is unjustified. Every individual has implanted within him the desire to understand his role in the existing order. He feels an inalienable right to find out his duties and privileges as a citizen of the universe. By the light of any new knowledge he is always certain to gain deeper insight into his position. The wise advice of our own great Emerson comes to mind: "Fear not the new generalization. Does the fact look crass and material, threatening to degrade thy theory of spirit? Resist it not: it goes to refine and raise thy theory of matter just as much."

What, then, are some of the larger points of view which are suggested by science today? In attempting a reply I can of course only offer a personal interpretation, inevitably reflecting the fact that I speak as a mathematician having some acquaintance with physics.

Let us observe in the first place that the universe presents antipodal aspects—the objective and the subjective, the impersonal and the personal. If we take the objective aspect as more fundamental we put our emphasis on the notion of reality; and if we start from the subjective, we prefer to speak of knowledge. In either case we are able to discern a kind of nature-mind spectrum; for there appears a roughly given hierarchy of five ascending levels—mathematical, physical, biological, psychological and social. Each level has its appropriate special language. The basic corresponding concepts are respectively: *number* at the mathematical level; *matter* at the physical level; *organism* at the biological level; *mind* at the psychological level; and *society* at the social level. If we choose to select one of these as somehow more real than the others, a great distortion arises in our point of view. For instance, if we regard the physical level as the most fundamental, we become materialists. But why make such an unnecessary choice? The languages of the various levels are essentially independent of one another, and the observed laws are best expressed in their own natural terms. Why mix up the levels of knowledge unnaturally? Does it clarify our idea of social justice to try to explain it in terms of the reactions between protons and electrons in the brain?

These considerations bring us to a first general point of view toward the levels of knowledge: It is desirable to accord reality in equal measure to all kinds of knowledge everywhere, and so to view the universe as broadly and impartially as possible.

Another very important observation is that in order to understand the various facts and their interrelations we must always use abstractions, that is, conceptual tools of a logical or mathematical nature. Contrary to opinions which prevailed until recently, any abstraction serves only limited specific ends. At best it will enable us to grasp more clearly some small fragment of reality. For example, by use of the abstraction of Euclidian geometry, and in that way alone, we understand the nature of space with a considerable degree of exactitude; and yet today scarcely any physicist would ascribe objective reality to space in itself. It has been Einstein more than any one else who has taught the scientific world the true role of Euclidian geometry by means of his theories of space-time and relativity. More generally, we have come to realize that our only approach to a better understanding of the world is by means of a widening succession of abstract ideas, each explaining imperfectly some aspect of the stupendous whole. This is a second synthesis deserving of especial emphasis.

Thirdly, I would state a fundamental truth about the social level, which in some sense is the highest level of all: The transcendent importance of love and goodwill in all human relationships is shown by their mighty beneficent effect upon the individual and upon society.

Thus, I have begun by presenting very briefly three important articles of my personal faith. These are not verifiable experimentally or strictly demonstrable, so that any one is free to agree or to disagree. Against my belief that the levels of knowledge are to be taken as equally real, one may set for instance an opposing belief that every fact is ultimately expressible in purely physical terms. If my position is natural for the mathematician with his abstract point of view, the other may be preferred by the tough-minded physicist, the biologist with mechanistic inclinations and the psychologist with a behavioristic outlook. The future will probably show that both of these beliefs are partly true and partly false.

Similarly, against my conviction that any particular abstraction is merely a useful tool enabling us to understand certain facts, some will contend that one particular abstraction will prove to be final and absolute. Here my attitude springs from an extensive acquaintance with mathematical abstractions and their numerous applications, whereas the theoretical physicist, for example, tends to believe that the ultimate theory of atomic structure is soon to be obtained.

Likewise some will declare that, much more than love and goodwill, it is devoted loyalty to the state which is important; and I can imagine that under certain conditions such an assertion might be justified.

It is my especial purpose to show how this phenomenon of faith arises inevitably in the mind of the scientist whenever he tries to evaluate technical conclusions in his special field. In doing so I shall discuss the role of intuition, reason and faith in science, first at the mathematical and physical levels, and then more briefly at the biological, psychological and social levels. This will lead me in conclusion to formulate two other items of my personal creed in the hope that they may be worthy of your attention.

By way of definition it must be indicated first what is meant by intuition. There are certain elementary notions and concepts which come spontaneously to the minds of all who observe, experiment with and reflect on a specified range of phenomena. Such generally accepted ideas or intuitions constitute the consensus of reaction of intelligent men to a definite part of the world of fact. John Stuart Mill has said, "The truths known by intuition are the original premises from which all others are inferred." It is in this sense that I shall refer to intuition. By reason I shall mean the rational superstructures which may be erected upon the basic intuitive ideas by means of deductive or inductive reasoning. These superstructures will also be accepted by all who are able to follow the sequence of logical steps involved. By faith I shall mean those heuristically valuable, more general points of view, which are beyond reason, and sometimes in apparent contradiction with one another, but which to the individual concerned seem of supreme importance as he endeavors to give his conclusions the widest possible scope.

It is clear that in this way we obtain a basic classification of knowledge into three easily distinguishable types. Let us consider the occurrence of these types at the various levels of knowledge.

By continual crude experimentation with classes of concrete objects, man has come gradually and inevitably into the possession of certain numerical ideas. In particular he has been led to think of the positive integral numbers 1, 2, 3… as entities which exist in almost the same sense as the objects themselves. This concept finds its realization in the designation of the integers by corresponding marks 1, 2, 3…. Such integers are found to be subject to certain simple arithmetic laws, and these laws are regarded as intuitively true.

The integers form the basis of a great part of mathematics. For it is found that with their aid one may construct fractions and, more generally, real and imaginary numbers. In the course of the centuries mathematicians have thus built, by processes of pure reason the elaborate structures of algebra, the theory of numbers and analysis. An extensive array of beautiful and useful theorems has been deduced.

Similarly in geometry—which in its origin may be regarded as the most elementary branch of physics we experiment with rigid material objects and arrive readily at the notions of idealized small rigid bodies or "points" and of idealized "lines" and "planes." Then we observe that certain postulates hold, such as the familiar ones of Euclid. By means of these postulates, which embody our intuitions, we are able by deductive reasoning to arrive at other geometrical theorems, including such results as the celebrated Pythagorean theorem which shows us in particular that a right triangle with legs of three units and four units in length has a hypotenuse of exactly five units in length. The vast mathematical domain called "geometry" has arisen from these elementary geometrical facts as a primary source.

There are many other abstract mathematical structures besides those just alluded to. In all cases it is found that they are made up of certain accepted intuitions (or postulates) and their logical consequences.

Now what I desire particularly to point out is that the mathematician goes far beyond such generally accepted clean-cut assumptions and conclusions, in that he holds certain tacit beliefs and attitudes which scarcely ever find their way into the printed page. Yet these form none the less part of a considerable oral tradition. For instance, he believes in the existence of various infinite classes such as that made up of all the integers. He believes also that the whole body of strict logical thought called mathematics is self-consistent: in particular when he finds that the number B admits of diverse forms of expression, as, for example

$$\pi = 4\left(\frac{1}{1} - \frac{1}{3} + \frac{1}{5} - \frac{1}{7} + \cdots\right)$$

and

$$\pi = 2\sqrt{3}\left(\left(\frac{1}{1}\right)\left(\frac{1}{1}\right) - \left(\frac{1}{3}\right)\left(\frac{1}{3}\right) + \left(\frac{1}{5}\right)\left(\frac{1}{9}\right) - \cdots\right)$$

and he feels absolutely certain that if the unending calculations could be fully carried out, the results would be exactly the same in all eases. (Note. Mathematicians will recognize these as the values of arctan x for $x = 1$ and $x = 1/\sqrt{3}$.) Furthermore, when he recalls that in the past the most difficult mathematical questions have been ultimately answered, he is inclined to believe with the great German mathematician, Hilbert, that every mathematical fact is provable. Besides all this, he attributes certain values to his results and their mathematical demonstrations: some theories seem important; some proofs are regarded as elegant, others as profound or original, etc.

Such somewhat vague ideas illustrate what I would call mathematical faith. Nearly all the greatest mathematicians have been led to take points of view falling in this broad category, and have attached the deepest significance to them.

What I wish to emphasize concerning this generally overlooked aspect of mathematical thought is that, on the one hand, the beliefs involved have been of the utmost heuristic importance as instruments of discovery, and on the other hand, when examined in detail, they generally turn out to involve ideas which are held true or false, according to the specific definitions which may be subsequently adopted.

Suppose, for instance, that we turn to the first question of the existence of infinite classes. There was no hesitation about the unconditional acceptance of such classes until within a few decades, although a few, like the ancient Greek philosopher Zeno and the German algebraist Kronecker, profoundly distrusted the use of the infinite in mathematical reasoning. Today, however, due primarily to the theory of transfinite aggregates created by Georg Cantor about fifty years ago, mathematicians have come to realize that such an infinite class may exist in the so-called "idealistic" sense but not in the sense of explicit constructibility. Thus the class of all collections of positive numbers less than one exists in the idealistic sense, but not in the alternative, more concrete sense.

A similar situation has arisen in the detailed study of the self-consistency of mathematics. It has appeared that very limited parts of mathematics can be proved self-consistent. But such a general assertion as that "the whole of mathematics is self-consistent" would be considered today not to be sufficiently precise; and each time that the proof of self-consistency is extended further, a definite logical price has to be paid in that certain so-called metamathematical ideas are tacitly employed, which need themselves to be investigated in the same respect. For instance, work prior to the *Principia Mathematica* by Whitehead and Russell (1910) showed that if the notion of class was not restricted, certain logical paradoxes would inevitably result. For this reason a theory of the "hierarchy of types" was devised by them, which limited the notion of class and so avoided the apparent inconsistencies. We are thus entitled either to say that mathematics as of the year 1900 was self-consistent or was not, according to the point of view which is adopted. In any case the belief in question has led us to a much deeper insight into the nature of logic.

With regard to the unlimited power of mathematical demonstration, it has been recently proved by the Austrian mathematician Gödel that, if we restrict ourselves to reasoning of an ordinary type, there exist explicit "undecidable" theorems, while from a higher metamathematical point of view such a theorem might be demonstrable. Hence Hilbert's affirmation is in one sense false. But despite this fact the open question on which he focused attention is much better understood than ever before.

Likewise in the question of value in mathematics, such as the importance of theories, or the elegance, profundity and originality of proofs, it is clear that these obscure ideas depend in large measure upon the momentary state of the science. Thus the theory of functions of an imaginary variable and classical geometry were regarded as extremely important a quarter of a century ago; while today the theory of functions of real variables and the basic kind of geometry called analysis situs have respectively displaced these subjects in general mathematical esteem. It would be hard to explain adequately the reasons for this change, but the increasing role of discontinuous quantity in physical theory and the relativistic point of view towards space and time have certainly been contributing factors.

An excellent instance of the power of individual mathematical faith in bringing about creative advance has been afforded by an American mathematician, the late Eliakim Hastings Moore, past president of this Association. Moore was a thoroughgoing abstractionist who believed that mathematics itself should be reorganized from a still higher point of view, by the dissection of essential common parts out of apparently different abstract fields. His point of view was strongly confirmed by the analytic work of Hilbert and Erhardt Schmidt near the beginning of this century. And so Moore was led to create his "General Analysis" in 1906. This aimed to embody his conviction that "The existence of analogies between the central features of various theories implies the existence of a general theory which underlies the particular theories, and unifies them with respect to these central features."

As time has elapsed, the deep truth of Moore's contention has been amply sustained. Indeed one of the most active schools of contemporaneous mathematical thought follows the higher abstract point of view adopted by Moore. But it has been found necessary to modify Moore's program, in that, instead of a single "General Analysis" serving as an *omnium gatherum,* it has been desirable to employ a few typical forms. In this way his faith in the power of higher abstraction has been largely and yet not fully justified.

A good many mathematicians are seriously hampered by lack of the ardent positive faith which Moore showed. This type of deficiency is generally due to a strong development of purely critical powers and to overspecialization. Several times I have observed this lack in myself, only to be counteracted by definite effort. For example, I did not make active use of the fundamental integral of Lebesgue for a long time, and so was prevented from pursuing to their natural conclusion certain ideas which finally led me to establish the basic "ergodic theorem" in 1931. Here I was finally converted, as it were, to the use of this tool by the important advances of Koopman and von Neumann, and in particular by the latter's proof of the "mean ergodic theorem." It is worthy of note that the related ergodic hypothesis goes back in its origin to the physicists Boltzmann and Maxwell.

Let us turn next to the physical level where the corresponding situation is at least equally interesting.

If we accept the ordinary conceptions of space and time, which seem destined always to play a basic role in workaday physics, we find that the simplest physical ideas are those which arise through the manipulation of massive bodies. As these ideas have become clarified, they have been given abstract formulation in terms of such concepts as those of mass, force, etc. Newton's celebrated three fundamental laws of motion embody the final form of the refined intuitions thus arrived at. With these as a basis and the acceptance of certain further special observed laws, one may deduce by mathematical reasoning the theory of mechanics as applied, for example, in the solar system

Similarly, through experimentation with electrified bodies, electric currents, magnets, etc., there was developed by Faraday the intuitive ideas of electromagnetic lines of force which are now generally accepted. Later Maxwell incorporated these ideas in the appropriate electromagnetic equations. Upon this basis all classical electro-magnetic theory has been logically constructed. Furthermore, by means of the identification of the light wave and the electro-magnetic wave, due to Maxwell, an adequate theory of light has been obtained.

Thus we see the important role which intuition and reason have played in two fundamental branches of physics—mechanics and electro-magnetism. A cursory survey of the various other branches of the subject would show that a similar situation holds throughout, except in the rapid developments of quantum mechanics during the last decade or so. In this strange theory the physicist begins indeed with a planetary model of the atom, reminiscent of Niels Bohr's earlier theory. But a flying leap is made from this temporary scaffolding to what is thenceforth regarded as the only basic reality—the wave equations of Schrödinger and better still, of Dirac. Once having arrived at these mathematical equations, the physical theorist proceeds to show how he can predict innumerable facts previously out of his range by use of this arbitrary *ad hoc* machinery. The process involved somehow reminds me of a record of a sea voyage made through a fog. I cannot but anticipate that a more intuitive and natural approach to essentially the same results will be found later on. An analogous earlier instance in physics is perhaps to be found in the unmotivated theory of cycles and epicycles entertained by the ancient astronomers. This explained the motions of the heavenly bodies with considerable success, but was destined to be completely displaced by the intuitively reasonable, gravitational theory of Newton.

The fact remains, however, that the recent development of quantum mechanics forms one of the most astounding and important chapters of all theoretical physics.

It is interesting to recall how this great advance came about through the faith of the German physicist Planck at the outset of the present century. His direct experience with the phenomena of radiation had led him to believe that there were discontinuous processes at work, not to be explained by any modification of the timeworn classical theories, and so he was led to formulate his celebrated quantum hypothesis in 1900. It was this daring concept of Planck, more than anything else, that has freed the minds of physicists from the shackles of too conventional thinking about atomic phenomena, and so has made possible the quantum-mechanical quest of which the end is not yet in sight.

There has always been an abundance of faith among the physicists. Everyone knows how Newton and others have found confirmation even for their religious beliefs in the lawful character of physical phenomena. It is not hard to understand why the tendency towards dogmatic affirmation among the physicists has been stronger than among the mathematicians. For the physicist with considerable justice feels that he is exploring the mysteries of the only actual and very exciting universe; whereas the mathematician often appears to live in a purely mental world of his own artificial construction. A good illustration of this tendency of the physicists is afforded by their changing attitudes towards the wave theory versus the corpuscular theory of light. Over a considerable period the corpuscular theory of Newton held sway; then this was displaced by the wave theory of Huyghens, the Dutch physicist; and nowadays a kind of vague, uncertain union of the two is generally accepted.

In this connection it is especially interesting to recall the scientific beliefs to which Faraday was led in his fundamental work on electricity and magnetism. From his experimental results in this field, he saw that there was obeyed here as elsewhere the law which he called the "conservation of force" and which we today would call the "conservation of energy." He saw that this energy was localized in space, and he could only conceive of it as being propagated in time; and so he was led to the belief that electro-magnetic energy is also propagated with finite velocity. Thus in an article, "On the Conservation of Force," published in 1857, he expressed himself as follows: "The progress of the strict science of modern times has tended more and more to produce the conviction that 'force [energy] can neither be created or destroyed ...; "*time* is growing up daily into importance as an element in the exercise of force; to inquire, therefore, whether power acting either at sensible or insensible distances, always acts in *time* is not to be metaphysical." By way of justification of the rather mathematical direction in these thoughts, Faraday said further, "I do not perceive that a mathematical mind, simply as such, has any advantage over an equally acute mind not mathematical..."; "it could not of itself discover dynamical electricity nor electromagnetism nor even magneto-electricity, or even suggest them." But the achievements of the more mathematical Maxwell were later to show that Faraday had underestimated the power of pure reason.

It is thus clear that through an act of faith Faraday attained to a kind of deeper insight; for the existence of the electro-magnetic wave has long since been established experimentally. However, the beliefs of Faraday in this connection cannot be regarded as absolutely true, since according to present-day conceptions the notion of energy which he accepted is only roughly valid as a statistical approximation. Nevertheless, Faraday certainly penetrated more into the nature of electrical and magnetic phenomena than any of his contemporaries; and it is difficult to see how, with the limited mathematical and physical knowledge at his disposal, he could have gone any further in the way of prophetic conjecture.

The intimate relation between philosophical-scientific points of view and actual advances in theoretical physics has been admirably illustrated by Einstein's gravitational theory of 1915. Taking as his starting point the bold but reasonable hypotheses that matter must condition space and time, and that, in parts of space remote from matter, elementary particles move with uniform velocity in a straight line, he arrived at his field equations as the most elegant mathematical embodiment of these ideas. Thus there was obtained a quasi-geometrical theory of gravitation which in certain respects is more natural than the celebrated theory of Newton, while the predicted differences, although excessively minute, are in favor of the new theory. But Einstein's theory cannot be regarded as true in any absolute sense, since it gives us at best a partial, highly idealized view of the physical universe.

It is hardly too much to say that, since the beginning of the present century, the main advances in theoretical physics have been the outcome of a similar kind of mathematical guesswork, in which, however, the mathematician himself has taken little or no part! The guessing of the physical theorist is guided almost entirely by considerations of subtle mathematical analogy.

This peculiar situation has led naturally enough to the feeling that pure mathematics almost suffices without much recourse to the results obtained in the physical laboratory. Sir Arthur Eddington has embodied the extreme point of view in his recent book, *The Relativity Theory of Protons and Electrons,* thus taking a position antipodal to that of Faraday. Eddington says: "Unless the structure of the nucleus has a surprise in store for us, the conclusion seems plain—there is nothing in the whole system of laws of physics that can not be deduced unambiguously from epistemological considerations. An intelligence, unacquainted with our universe but acquainted with the system of thought by which the human mind interprets to itself the content of its sensory experience, should be able to attain all the knowledge of physics that we have attained by experiment.... For example, he would infer the existence and properties of radium, but not the dimensions of the earth."

I would comment upon this mystical conjecture of Eddington as follows: It is no doubt partially true that in some respects we need the laboratory less than we did before, due to the fact that we live surrounded by all manner of scientific instruments and machines, with whose properties we have become acquainted. In other words, we live in a transformed world which is a kind of huge laboratory. Yet I doubt whether any individual, however intelligent, who was not acquainted with such instruments and machines, would be able, through analysis of ordinary sensory experience, to go very far. On the other hand, I would agree with Eddington that the starting point from which known physical laws may be deduced is likely to depend on only a few intuitive ideas; and perhaps a sufficiently powerful mathematical intelligence would realize that the facts of sensory experience could only be simply explained in this way.

An equally remarkable conjecture was expressed by Dr. Charles Darwin in a vice-presidential address, "Logic and Probability in Physics," before the British Association last summer. In this address he said, "The new physics has definitely shown that nature has no sharp edges, and if there is a slight fuzziness inherent in absolutely all the facts of the world, then we must be wrong if we attempt to draw a picture in hard outline. In the old days it looked as if the world had hard outlines, and the old logic was the appropriate machinery for its discussion." He therefore suggested "that some day a real synthesis of logic will be made" leading to "a new reformed principle of reasoning."

Here I can agree with Darwin to the extent of admitting that there always exists a metamathematical fringe in logic. But it seems obvious that in logic there has been a record of continual advance by critical and profound diversification rather than by any essential alteration of point of view.

In my own limited experience in mathematical physics I have also seen how natural it is to take a positive attitude on open questions. Thus a good many years ago I showed mathematically that mere spatial symmetry about a center necessitates a *static* gravitational field. This led me to believe that the Einstein field equations were probably too inelastic to fit the facts, but I did not put forth this opinion. Shortly afterwards Lemaître, in trying to explain the expanding (non-static) stellar universe found it necessary to modify the field equations, in part because of my result; and so my belief was to this extent justified.

Again, I have had during the last few years a feeling that a conceptual space-tiine model for quantum mechanics is likely to be found, although theoretical physicists would in general disagree. Nevertheless, my faith is so strong that my recent researches lie principally in this direction. I have already found interesting results, and am confident that these efforts will not be wasted, since the possibilities of the conceptual approach need to be more carefully explored.

In ending these remarks about the role of intuition, reason, and particularly of faith, at the physical level, it is to be observed that the physicist as such systematically ignores the phenomena of life, for it is dead and not living matter with which he concerns himself in his laboratory.

All in all, it is a faith in the uniformity of nature which remains the guiding star of the physicist just as for the mathematician it is a faith in the self-consistency of all mathematical abstractions, although these faiths are more sophisticated than ever before. The minds of both are tinged with an unwavering belief in the supreme importance of their own fields. The mathematician affirms with Descartes, *omnia apud me mathematica fiunt*—with me everything turns into mathematics; by this he means that all permanent forms of thought are mathematical. The physicist on his part is apt to think that there is no reality essentially other than physical reality, so that life itself is finally to be fully described in physical terms.

Although I have no especial acquaintance with the biological, psychological or social domains, it seems clear to me that a similar situation prevails in them. In the biological field the intuitions upon which one depends are those associated with the concept of the organism and its evolution. These intuitions can not be formulated conclusively and completely in simple postulates, as is possible at the mathematical and physical levels. It is rather through an acquaintance with an immense array of interrelated, analogous facts that the biologist finds himself able to deal with novel situations. By means of the geological record on the one hand and the results obtained in the field and laboratory on the other, he acquires a better and better understanding. His principal weapon is always inductive reasoning. It seems certain that a deductive treatment of biology is at least very remote and if ever accomplished will be utterly different from anything which we can imagine today. There are, however, a few special fields like the theory of heredity, in which a considerable mathematical structure has been developed. In this theory, by means of the "chromosomes" and their corresponding abstract "genes," it has been possible to explain a complicated array of facts.

The faith of the biologist generally tends in the direction of a mechanistic theory of life or of some opposing vitalistic theory. In fact, he is forced to employ the principle of physical causation in his efforts to understand biological phenomena and does not yet know of definite limitations in its use. Recently there has been some indication of a return to vitalism, so that once more a considerable group of biologists are convinced that not all the phenomena of living matter are to be accounted for by ordinary physical and chemical laws. The controversy involved has long been a burning one, and accordingly one naturally suspects that the question is really meaningless. In any case, however, special mechanistic hypotheses have so far pointed the way to new creative advances.

It is interesting to remark that the insufficiency of a rigorously deterministic theory of the living organism admits almost of mathematical demonstration in the following manner. A genuinely mechanistic universe would have to be free of any infinite factors. For example, if one accepts a simple Newtonian theory, there might be reaching the earth from infinite space unknown quantities of matter and energy, so as to change arbitrarily the course of events upon the earth. But in any completely mechanistic system, free of such infinite factors, it is not difficult to prove that there will necessarily be a kind of eternal Nietzchean recurrence. For instance, we are here together this evening considering a particular topic. The strict adherence to the deterministic point of view would entail the consequence that in the eons yet to come this same scene will be re-enacted infinitely often. I submit that this is dramatically improbable!

Recent advances in the chemical knowledge of large organic molecules seem to indicate an innate hospitality of actual matter toward the evolution of the living organism. In this way a plausible genetic account of the origin of life is suggested, which, however, can scarcely be called mechanistic. It begins to seem possible that we are on the verge of further refinements in our concept of matter, such as Emerson anticipated in the quotation made above.

The situation at the psychological level is even less amenable to precise treatment. All of us have a lifelong experience with ourselves and other human beings. This automatically gives rise to a vast complex of intuitive psychological notions. We all are aware of course that there are concomitant physiological processes going on in the body, nervous system and brain. Now it is the business of the professional psychologist to give exact definition and interpretation to these crude ideas; and he finds his greatest illumination in the facts of abnormal psychology, with which most of us are unacquainted. However, in the case of either layman or professional the processes of reasoning are mainly by analogy. Even the psychiatrist, familiar with many concrete cases, must treat each new patient by the inductive method. There are too many psychological intuitions and too few exact laws for any imposing edifice of pure reason to be erected.

In certain restricted psychological domains, formalization is to some extent possible. Thus I have ventured to formulate a theory of "esthetic measure," by explicit numeration and weighting of esthetic factors. This aims to explain certain simple esthetic facts in our enjoyment of visual and auditory forms. The theory has been to some extent substantiated by experiments made at Harvard and elsewhere. But in any case, no matter how successful the theory might prove, it would be wholly absurd to try to set up an elaborate logical structure on the basis of the fairly arbitrary and inexact assumptions involved. Generally speaking, as we proceed from the more objective to the more subjective levels of thought, we find that elaborate logical structures seem to be of less and less utility.

The basic belief of the professional psychologist is in the completeness of the physiological accompaniment of every psychical fact; and he formalizes the observed facts by means of the parallelism. But there is a conflict between this attitude of the technician towards mind, for whom the individual is a complex of neurally characterized components, and that of the ordinary man—equally an expert though of a different kind—who sees all sorts of permanent values in personality, not adequately characterized in neural terms. The second attitude leads nearly all of us to have deep affections and abiding personal loyalties, whether or not we are psychologists!

Here again I think that these apparently opposing points of view are both more or less true; and I incline all the more to this opinion because of my conviction that as yet we know relatively little about the phenomena of personality. For it seems certain to me that the extent of hidden organization in our universe is infinite, outside as well as inside of space and time; such a conviction is very natural to a mathematician, since the three ordinary spatial dimensions and the single temporal dimension are for him only particular instances of infinitely many other conceivable dimensions: If this be true, any broad conclusions concerning the nature of personality would seem altogether premature.

At the social level the most serviceable intuitive ideas cluster around the concept of societal evolution. It is of course the comparative study of human institutions which furnishes the principal interest. The analogy between forms of society and evolving organisms is a deep-lying one. Here again the useful logical structure which can be built around the very complicated facts is exceedingly simple. Even in such a formalized field as ethics, dealing with the behavior of the individual as a member of society, logic plays an almost negligible role.

Belief here seems to gather principally around the idea of societal progress. Progress—or its non-existence—serves as our fundamental tenet. Some believe that society can improve indefinitely, tending toward a perfect society. Such a belief is of course a fundamental one in most religious systems. Others find this idea too naive. They stress the gregarious instinct in man and tend to think of societal changes as taking place in various directions strongly conditioned by changing physical environment. All would admit, however, that without the concept of dynamical social processes, social theorizing would be stale and unprofitable.

Let us turn now to consider some further conclusions, towards which this brief survey of intuition, reason and faith at the various levels seems to point.

As far as intuition and reason are concerned, these are the common property of all competent individuals. The narrow, closely articulated chains of deductive reasoning serviceable at the earlier levels are more and more replaced by loose webs of inductive reasoning at the later levels, as we pass from the objective to the subjective. At the same time the basic intuitions change from the simple and precise types employed in mathematics and physics to the increasingly complicated and diverse forms characteristic of biological, psychological and social phenomena.

However, it is just as necessary to clarify and to formalize our knowledge at these later levels as at the earlier ones. The processes of systematic reasoning, whether inductive or deductive, have always a definite prophylactic value, and in particular enable us to avoid the dangers of prejudiced and intolerant points of view. It may be observed in passing that the careful application of impartial thoroughgoing analysis is as important for everyday living as it is in the study and the laboratory.

The striving for rational comprehension is one of the noblest attributes of man. In his agelong difficult struggle he has been able to secure greater freedom only through a better technical mastery of his environment. No other method of liberation has been vouchsafed to him. But this increased mastery has brought with it automatically new intellectual responsibilities and a more complex way of life. In consequence, unforeseen and threatening dangers arise from time to time; and there is thus imposed on him the necessity to advance still further, which is today more urgent than ever before.

A new injunction has been laid upon the spirit of man, to know and to understand ever more broadly and deeply.

Now along with the increase in scientific knowledge there appear certain crudely expressed, deeper insights, not completely true or false, some in opposition to others, but all supremely valuable nevertheless. These are embodied in beliefs which seem the inevitable accompaniment of all creative thought.

Thus in the daring effort of the scientist to extend knowledge as far as possible, there arises an aura of faith. It is this spontaneous faith which furnishes the most powerful incentive and is the best guide to further progress.

Such are some of the very general points of view to which a considerable mathematical and scientific experience has led me. If they are worthy of serious attention it is not because of their novelty, but rather because in their aggregate they rise above the details of the numerous specialized fields of knowledge and sustain the scientist in his unceasing and ardent search after truth.

Doubtless many of you are ready to ask the ever more insistent question: If science has thus profoundly modified the general outlook and way of life of mankind, is it not the especial duty of such an association as ours to point out constructive remedies for the ensuing maladjustments? In the "Part II: Science and Warfare" of his admirable address as president of the British Association last August Lord Rayleigh closed by expressing the hope that our two associations could cooperate in such a way as to "bear useful if modest fruit in promoting international amity." In this hope all of us will deeply concur. The presence of Sir Richard Gregory with us at the Richmond meeting is the first token of the projected closer relation between the parent British Association and ourselves. It is much to be desired that this action will encourage further unification of the whole scientific world. I am sure that practically all our joint membership would agree with me that it is the wider diffusion of "the steady light of scientific truth" which holds out most hope of a better understanding among men.

"Medicine makes people ill, mathematics makes them sad, and theology makes them sinful."
—Martin Luther (1483–1546)

Logic and the Understanding of Nature

David Hilbert

David Hilbert (1862–1943), one of the most influential and creative mathematicians of the 20th century, was born in the city of Königsberg, then the capital of East Prussia, now in Russia and renamed Kaliningrad. Königsberg is also the city of Immanuel Kant who exerted considerable influence on the citizens of Königsberg. Its main mathematical claim to fame, however, stems from the seven bridges that cross the river and join an island in the river. These seven bridges gave rise to a famous problem that was solved by the celebrated Euler; its solution gave birth to the discipline of topology.

Hilbert's father was a judge but it is thought that Hilbert derived his mathematical talent from his mother, who is reputed to have been fascinated by prime numbers.

Although the family were not church-goers, it is likely that they had Pietist roots. One of Hilbert's antecedents for example was named "Fear God Live Right" reminiscent of the Puritan "Praise God" Barebones. The Pietist movement rebelled against the established church and stressed repentance, faith, and sanctification. Its analogue in England was the Puritan movement.

Hilbert entered the classical gymnasium where he met Hermann Minkowski who became a lifelong friend until the latter's premature death. Hilbert exhibited an extraordinary talent for mathematics and received the degree of doctor of philosophy in 1884. He then traveled widely and met a galaxy of some of the most noted mathematicians of the period—Weierstrass, Kummer, Kronecker, Klein, Helmholtz, Weber, Poincaré and Jordan—the last two in Paris.

In 1895 he was appointed to a professorship in Göttingen, the university of Gauss, Dirichlet, and Riemann. He remained there the rest of his life.

He achieved his first mathematical triumph with a proof of the finite basis theorem in algebra and became widely known for his treatise in number theory–a work which fascinated Hermann Weyl. These achievements were matched by equally important and influential works throughout his life. Noteworthy is his work on the foundations of geometry and his investigations in logic. His influence was especially felt by the number of students he supervised, one of whom was Hermann Weyl. Several came from the United States to study with the master.

In 1892, he married Kathe Jerosch, his second cousin, who became his life-long friend and companion. They had one son Franz.

Göttingen was a world center of mathematical activity; there was a current saying among the inhabitants "outside Göttingen there is no life." He became well-known in a wider circle when, in 1900, he gave a lecture to the International Congress of Mathematicians in Paris entitled "Mathematical Problems." In this lecture, he set forth a set of 23 problems whose solution would, in his view, contribute significantly to progress in mathematics. These problems are still the focus of considerable interest in the mathematical community. Many have now been solved.

His creativity continued without cease and, over the years he was showered with honors. This idyllic life was disrupted by the accession to power of the National Socialists in Germany. The expulsion or murder of "non-Aryan" mathematicians was a source of great grief to Hilbert and his wife but in their twilight years they felt powerless to do much about it.

Hilbert died in Göttingen in 1943.

Editor's Preface

This article which appeared in 1930 was originally given as a lecture; it is a sort of culmination of Hilbert's ideas that were developed and matured over a number of years.

The principal theme is the question as to the relative roles of observation and thought in the acquisition of knowledge. The present (1930) status of which, he says, is particularly favorable to illuminate this question. He relies heavily on the wonderful and abundant discoveries in all the sciences that were made in the first decades of the twentieth century. In his discussion, Hilbert says that axiomatics and logical deduction have an essential role to play. The discipline of formal logic was essentially created in Hilbert's lifetime and his own contributions were pivotal. The kernel of Hilbert's question can be formulated as the question, raised in the past by philosophers, scientists, and theologians, viz., "Where does reality reside?".

He notes, however, that besides deduction and observation, there is historically another source of knowledge—that of Kantian "a priori." Kant introduced this concept in the *Critique of Pure Reason.* By a priori he meant that certain facts and concepts are inherent in nature and are independent of experience. This contrasts with the empiricists who maintain that beliefs are to be accepted only if they are confirmed by experience. Hilbert asserts that Kant and his disciples had greatly overstated their case. He bases his rebuttal partly on an example, viz., that of Euclidean geometry that was pivotal in Kant's discourse. Gauss had already disputed Kant's claim, since he had already constructed a model for non-Euclidean geometry—a model that did not assume the parallel postulate [this postulate was a crucial assumption in Euclid's development of geometry]. Gauss never published anything on the subject for fear of the "cries of the Boeotians" The axioms of Euclid were deficient in another respect since they said nothing about betweenness, or about "inside" and "outside," etc., and one of Hilbert's important early contributions was the recognition of those additional axioms needed to vindicate the geometry of Euclid. He claimed that no set of axioms of Euclid is completely satisfactory until the concepts of "point," "line," and "plane" could be replaced with arbitrary terms such as "table," "chair," and "mug." In other words, Euclidean geometry is pure thought and deductions follow entirely from the rules of logic. This served as a model for other examples.

Having helped to establish logic as a powerful discipline, Hilbert undertook to use this instrument to analyze further the relation between thought and experience. His methodology is essentially that espoused by the "logical empiricists," a movement that gained some popularity in the twenties. Mathematics is in the midst of all this he claims; it is the handmaiden of scientific investigations and is the agent between observation and thought. He ends on a note of unalloyed optimism. He rejects the pessimism of Auguste

Comte who endeavored to pose an unanswerable question. He concludes with what is, by now, his well-known slogan, "We must know, we shall know." It is more emphatic in the original German: Wir müssen wissen, wir werden wissen.

The understanding of life and nature is one of our foremost tasks. All human endeavors and desires move toward that goal and ever increasing success is accordingly achieved. We have obtained richer and deeper knowledge about nature in the past decades than we had in previous centuries. Today we shall use this advantageous position, following our theme, to deal with an old philosophical problem, viz. the controversial question concerning the relative importance of thought, on the one hand, and experience on the other, in our search for knowledge. This old question is basic since to answer it means essentially to state what the nature of our science is, and in what sense all the knowledge that we acquire in the domain of science, is truth.

Without presumptuousness toward the old philosophers and researchers, we can at present, with more certainty than before, reason with more confidence toward a resolution of this question; and that on two grounds. The first is the abovementioned rapid pace with which science develops today.

The significant discoveries of earlier times, from Copernicus, Kepler, Galileo, Newton till Maxwell, are spread out in large intervals spanning four centuries. The recent period begins with the discovery of Hertz waves. Then follow one after another: Roentgen discovers his rays, Curie radioactivity, Plank lays the foundation for quantum theory. And in the most recent times, discoveries of recent and surprising connections have streamed forth, so that the richness of the images becomes almost unsettling. Rutherford's theory of radioactivity, Einstein's $h\nu$ law, Bohr's clarification of the spectrum, Mosely's enumeration of the elements, Einstein's relativity theory, Rutherford's analysis of nitrogen, Bohr's structure of the elements, Aston's theory of isotopes.

So we witness in physics alone, an unbroken series of discoveries—and what discoveries! In this vastness not a single one of the attainments of older times is to be found; moreover discoveries at present are more compact and indeed more internal and more varied than the earlier ones. Hence they exhibit consistent theory and practice, thought and experiment, all of which are intimately intertwined. Now proceeds theory, now experiment, mutually validating, supplementing and motivating. Similar observations apply to chemistry, astronomy and the biological sciences.

We have the advantage over the old philosophers of experiencing a large number of such discoveries and have learned the genesis of these new settings. Thus there were among the new discoveries, many that forced a change or abandonment of the deeply rooted perceptions and ideas of the older sciences. Let us think, for example, of the new concepts of relativity theory or about the decomposition of the chemical elements; preconceptions were set aside, which earlier, no one in general would have undertaken.

But yet a second instance comes today: the solution of various old philosophical problems as a bonus. We have arrived today at a heretofore unattained height, not only in the technique of experiments and the art of building theoretical-physical structures, but also the complementary study, namely the science of logic, has been significantly advanced. There is today a general method for the theoretical treatment of scientific questions, which in all cases helps to clarify the articulation of the problem and helps to prepare it for solution. This namely is the axiomatic method.

What sort of thing is this often mentioned axiomatic method? The basic idea depends on the fact that, even in comprehensive domains of knowledge, only a few statements—called axioms—usually suffice. But even with this remark its significance is not fully appreciated. Examples can quickly clarify the axiomatic method for us. The oldest and best known of the axiomatic method is the geometry of Euclid. I should prefer however to clarify the axiomatic method very briefly with an especially amazing example from modern biology.

Drosophilia is a small fly, but our interest in it is very great; it is the subject matter of the most extensive, the most careful, and most successful investigations into breeding. This fly is usually grey, red-eyed, spotless, split-winged, or clump-winged. There are however, flies with abnormal characteristics: instead of grey, they are yellow, instead of red-eyed, they are white eyed, etc. Usually these five characteristics are coupled, i.e., when a fly is yellow, then it is white eyed and spotted split-winged and clump-winged. And when it is clump-winged, it is yellow and white eyed. From this inherent coupling come however, by cross-breeding among the descendants, a fewer number of abnormalities and indeed in a constant proportion. From the numbers that one thereby finds experimentally, arise the linear Euclidean axioms of congruence and the axioms of the geometric concept of "between"; and so arise the laws of heredity, as an application of the axioms of linear congruence, i.e., the elementary theorems about translation of line segments; the relation is so simple and precise—as well as wonderful, that no audacious phantasy could have invented it.

A further example of the axiomatic method in another domain is the following.

In our theoretical sciences, we are used to the application of formal thought processes and abstract methods. The axiomatic method belongs to logic. In wider circles, the word logic connotes a very dull and difficult thing. Today the science of logic has become easy to understand and is very interesting. For example, we have already seen that in daily life new methods and the building of concepts have come into use; these demand a whole level of abstraction and can be understood only through unknown applications of the axiomatic method, e.g., through the general process of negation and in particular through the concept of "infinite."What the concept of "infinite" entails is something we must make clear; that "infinite" has no clear significance and without closer investigation has no meaning, since in general there exist only finite objects. There is no infinite oscillation, no infinitely rapid propagation of force or energy. Moreover energy itself has a discrete nature and exists only in quanta. There exists in general no continuum that can be divided infinitely often. Light has an atomic structure just as energy does. Even space, as I definitely believe, has a finite scope and the astronomers could at one time give the dimensions of the universe—length, breadth, and width. We also have cases of very large numbers such as the distance of the stars in kilometers or the number of essentially different possible chess games, and since negation has a surprising status, so also is finiteness or infiniteness a gigantic abstraction, attainable only through the conscious or unconscious application of the axiomatic method. This perception of the infinite, which I have established in exhaustive detail, solves a series of basic questions, in particular the Kantian antinomy concerning space and the possibility of unbounded division, and thus my investigations solved the difficulties which, at one time, interposed themselves.

When we turn toward our problem viz, how nature and thought are interrelated, we shall consider three principal points. The first concerns the problem already mentioned viz. that of the infinite. We saw that the infinite is never realized. It is available neither in nature, nor as a basis in our thinking, without exercising special precaution. Here I have already discovered an important parallel between nature and thought, a fundamental agreement between experience and theory.

Yet we take another parallel as true—our thought arises from unity and seeks to build unity; we observe the unity of substance in matter and we determine, in general, the unity of nature's laws. Consequently nature, in reality, comes to us without resistence, as though it were ready to divulge its secrets. The thin distribution of matter in the universe makes possible the discovery and the more precise determination of Newton's laws. Michelson, despite the great speed of light and the falsity of the addition theorem, could determine the velocity of light, since our earth makes its orbit around the sun. The planet Mercury does us the favor of carrying out its perihelion motion so that we can verify the theory of Einstein. And the fixed star path goes around the sun in such a way that its deviation can be observed.

The appearance of what we call intrinsic harmony is also striking, in a sense other than that used by Leibniz, that it is an embodiment and realization of mathematical thought. The old examples of this are the intersections of the sphere, which were studied long before anyone guessed that our planets or indeed electrons moved in such paths. But the most noble and most wonderful example of intrinsic harmony is the famous Einstein relativity theory. Here the differential equations of gravitational potential are uniquely established entirely through the general progress of invariance, in combination with principles of the greatest simplicity, and these determinations would have been impossible without the deep and difficult mathematical investigations of Riemann; these had been in existence long before. In the most recent times, cases are being considered that are most important and at the center of interest for mathematical theories; they are also noteworthy in physics. I had developed the theory of infinitely many variables as being of purely mathematical interest and subsequently applied it to the notion of spectral analysis, without having any idea that a little later this would be realized in the real spectra of physics.

We can only understand this agreement between nature and thought, between experiment and theory, if we take into consideration the formal component of both sides of nature and our understanding, and the mechanism on which it depends. The mathematical process of analysis gives us, or so it appears, the focus and footings to which matter in the real world, as well as thought in the world of the mind, withdraw and cede control and direction.

Meanwhile this established harmony does not yet create a connection between nature and thought and does not yet reveal the deepest secrets of our problem. To grasp this, let us put in our purview the totality of the physical-astronomical complex of knowledge. We notice then, in contemporary science, a point of view which goes far beyond the old questions and objectives of our science: it is the case that contemporary science, not only in the sense of classical mechanics that teaches, from current data, the future motion and determination of expected phenomena, but it shows also that the present existing state of the material of the earth and the universe is not accidental or arbitrary but follows from physical laws.

The most important evidence is the Bohr model of the atom, on which is built the world of the stars and, finally, the entire development of organic life. So it seems that the pursuit of this method must lead in reality to a system of laws of nature which are suitable in their total validity and then in fact require only thought, i.e., abstract deduction, to master all physical truth; therefore Hegel would have been right in the claim that all natural occurrences can be deduced from concepts. But this conclusion is not valid. What about the origin of the laws of the universe? How can we discover these? And who teaches us that they are suitable for reality? The answer is evident, that this exclusively makes our knowledge possible. In contrast to Hegel, we recognize that the universe's laws cannot be obtained in any other way except through experience. By the construction of a framework of the physical concepts, many speculative points of view cooperate: either the constructed laws, and the logical framework built out of these are valid, knowledge alone is capable of deciding. Sometimes an idea has its origin in pure thought as, for example, the idea of atomism in Democritus, while the existence of the atom was proved two millennia later through experimental physics. Sometimes an idea arises and compels the mind to a speculative point of view. So we are grateful for the ingenious stimulus of the Michelson experiment, that the deeply rooted conception of absolute time was overturned and finally the idea of general relativity of Einstein could be established.

Nevertheless those who deny that the laws of the universe arise from experience, claim that, apart from deduction, a third source of knowledge exists. There are in fact philosophers, and Kant is the classical representative of this point of view, who claim that besides logic and experience we have an "a priori" certain knowledge of reality. Now I claim that for the construction of the theoretical framework, certain "a priori" insights are necessary and that the perception of our knowledge always has such a basis. I believe also that mathematical knowledge eventually will be based on a certain kind of intuitive insight. Moreover, in the construction of number theory a certain intuitive justification is an a priori necessity. Therein lies the significance of the most general basic idea of Kant's epistemology; namely the philosophical problem of investigating every perceptual experience which is ascertained a priori and thereby to investigate the conditions of the possibility of all conceptual knowledge as well as every experience. I mean, this appears essentially in my investigations on the principles of mathematics. The "a priori" is thus no more nor less than a basic observation or the expression for certain essential prerequisites for thought and experience. But the limit on the one hand between that which we a priori possess, and on the other hand that for which experience is essential, should be seen in a way other than that of Kant; Kant had overestimated the role and scope of the a priori.

One could assume that in the time of Kant, the perception of time and space which they had, could in general and without restriction, be applied to reality, as for example, to our representation of number, of infinite series, and of magnitudes that we constantly use in mathematical and physical theories in a familiar manner. Then, in fact, the theory of time and space, in particular geometry and arithmetic, precede all knowledge of nature. But this point of view of Kant had already, and with complete justification, been abandoned by Riemann and Helmholtz even before the developments of physics forced it; since geometry is none other than a certain part of the total structure of physical concepts which maps the possible relations of moving bodies one to another in the world of real things. That there are moving rigid bodies and what the relations are among them is entirely a matter of experience. The theorem that the sum of the angles of a triangle amounts to two right angles and that the parallel axiom holds, as was recognized by Gauss, can only be verified or refuted by experiment. If we could prove, for example that the various facts which could be expressed through the congruence theorem, coincide with experience, that the sum of the angles of a triangle made of rigid rods is smaller than two right angles, no one could then imagine that the parallel axiom in the space of real bodies is valid.

Acceptance of the "a priori" certainty requires the exercise of the greatest caution since indeed, many of the earlier facts assumed to be a priori are no longer recognized as being valid. The most striking example is the concept of the absolute present. Although we are used to it, and from childhood assumed its validity, there is no absolute present since even in daily life it would entail only small distances and slow movements. If it were otherwise, then no one could come to the point of introducing the concept of absolute time. Yet such deep thinkers as Newton and Kant had never reached the point of doubting the absoluteness of time. The cautious Newton formulated this postulate as crudely as possible: absolute time flows uniformly in virtue of its nature, uniformly and without any relation to matter. Newton had thereby properly rejected any retreat or compromise, and Kant, the critical philosopher, showed himself uncritical, since he accepted Newton without further ado. Einstein was the first to free us from this misconception—it remains one of the most significant creations of the human mind—and the excessive and widely used "a priori" theory, could not lead strikingly to a contradiction despite progress in the physical sciences. The assumption of absolute time had as a consequence, among other things, the theorem on the addition of velocities, a theorem which could not refute the evidence and popular understanding—in fact it follows convincingly from various experiments in the domains of optics, astronomy, and electricity—that this theorem on the addition of velocities is not true; there is, in fact, another complicated law for the combination of two velocities. We can say that in recent times, the Gauss-Helmholtz intuitive substitute for the empirical nature of geometry is a more definite move toward scientific progress. It should be of concern to all philosophical speculations to use this as a basis. Indeed the Einstein gravitation theory makes it evident: geometry is not a branch of physics; in principle, geometric truths are not in any way or form other than physical. So for example, the Pythagorean theorem and the Newton law of attraction are related to one another since they are governed by the same fundamental physical concept, that is, potential. But there is more for those familiar with Einstein's gravitation theory: these two theorems, so seemingly different (and until now separated by time, the one already known in ancient times and in school, a theorem of elementary geometry, the other concerning the force of masses on one another), are not only of the same character but are part of one and the same general theorem.

The principal similarities between geometric and physical facts can hardly be treated more dramatically. To be sure, with the usual logical structure, and from childhood on, with our common daily and familiar experiences, arise the geometric and kinematic theorems of dynamics, and the fact, if one has forgotten, that they are, above all, experiences. We see thus: in the Kantian "a priori" theory, there are still anthropomorphic residues from which it must be freed, and after removal, only those a priori justifications remain which are based on mathematical knowledge; it is, in principle, what I have characterized as finite justification.

The tool which governs the mediation between theory and practice, between thought and observation is mathematics; it builds the bridge and carries more and more of the load. It thereby happens that the basis of our entire present day culture, in so far as it is based on investigations dealing with nature, can be found in mathematics. Galileo had remarked: Nature can only be understood by one who learns the language and symbols in which it speaks; this language however, is mathematics and its symbols are the mathematical forms. Kant coined the phrase: "I claim that in every particular science of nature, only that part can really be understood that contains mathematics." In fact we cannot master a scientific theory, until such time as we have revealed the mathematical kernel and exposed it completely. Without mathematics, contemporary astronomy and physics would be impossible; these sciences in their theoretical parts are solved with mathematics. These and the numerous further applications, whose constructs are attributed to mathematics, are such that they are enjoyed by a wider audience.

Despite this, mathematicians have denied that the applications are a measure of the value of mathematics. The prince of mathematicians, Gauss, who was certainly an applied mathematician par excellence as well, who played a leading part in the entire science, such as error theory, newly created geodesy; in addition, when after the astronomers had recently discovered the planet Ceres—a particularly important and interesting planet—and then lost it and could not find it again, Gauss conceived a new mathematical theory on the basis of which the location of Ceres was predicted; he invented the telegraph and many other practical things; he was of the same opinion. Pure number theory is that domain of mathematics, which up to the present has not found any applications. But number theory is exactly what Gauss called the queen of mathematics and is exalted by almost all great mathematicians. Gauss speaks of the magical enchantment which number theory, which outshines all other parts of mathematics, had given to the early mathematicians; it was their favorite subject; and its unconquered kingdom gave pleasure as well. Gauss testified, how in his earlier youth the kingdom of number theoretic investigations had so captivated him that he could not leave it alone. He eulogized Fermat, Euler, Lagrange and Legendre as men of incomparable stature, since they opened and showed the entrance to the sanctum of this heavenly science, and they showed what great wealth it possesses. The mathematicians such as Dirichlet, Hermite, Kronecker and Minkowski, both before and after Gauss, spoke enthusiastically in a similar way—then compared number theory to a lotus eater in that once you have taken of this food, you could no longer leave it alone.

Also Poincaré, the most brilliant mathematician of his generation, who was a significant physicist as well as astronomer, was of the same opinion. Poincaré turned once, with striking sharpness, against Tolstoi who had stated that the claim "science for the sake of science" is foolish. "Should we", said Tolstoi, "allow ourselves, by the choice of our occupation, to be led by the whim of our inquisitiveness? Would it not be better to make the decision according to its utility, i.e., according to our practical and moral demands? "Tolstoi is certainly peculiar; we mathematicians must reject such prosaic realism and unsympathetic utilitarianism." Poincaré expostulated against Tolstoi, that if one follows the recipe of Tolstoi, science would never have arisen. One has only to open one's eyes to see how, for example the achievements of industry would never have enlightened the world if these practical people alone had existed as well and if these achievements had not encouraged the opening of gates, they would never have thought of practical applications. We all share this same opinion.

Even our great Königsberger mathematician Jacobi thought so; Jacobi whose name stands beside that of Gauss and whose name today is mentioned in awe by every student of our faculty. As the famous Fourier once said, the principal aim of mathematics lies in the clarification of natural phenomena, and it was Jacobi who countered with some anguish. "A philosopher, as Fourier certainly was, should have known," so said Jacobi, "that the unique goal of science is the honor of the human spirit and from this point of view, a problem in number theory has just as much value as one which serves applications."

Whoever is enlightened by Jacobi's liberal way of thinking and his world view, will not descend to a reactionary and fruitless skepticism; he would not believe what today they prophesy, with philosophical demeanor and strident tone, viz. the decline of culture, and moreover, would not fall into a state of *Ignorabimus*. For the mathematician, there is no *Ignorabimus* and in my opinion none at all for the natural sciences. Once, the philosopher Comte posed the question of finding the chemical composition of the heavenly bodies. His object was to pose a certain unsolvable problem—which objective cannot succeed in science. A few years later this problem was solved by Kirchhoff and Bunsen using spectral analysis and today we can say that we can lay claim to the most distant star as the most important physical and chemical laboratory, one that we cannot find on earth. The true basis on which Comte could not succeed in finding an unsolvable problem, in my opinion, consists in the fact that there is no unsolvable problem at all. In place of the foolish *Ignorabimus* is, in contrast, our slogan:

We must know
We shall know.

"The union of the mathematician with the poet, fervor with measure, passion with correctness, this surely is the ideal." — William James (1842–1910)

The Cultural Basis of Mathematics

Raymond Louis Wilder

R.L. Wilder (1896–1982) was born and reared in Palmer, Massachusetts where he received his early education. The family was musical and he learned to play the cornet and appreciate all forms of music, especially classical.

He entered Brown University in 1914 and after a stint in the armed services in World War I he completed his bachelor's and master's degrees in actuarial sciences in 1920 and 1921 respectively. In the same year he married Una Maude Green and together they made their way to the University of Texas. Wilder intended to get a PhD in actuarial mathematics. Instead he came under the spell of R.L. Moore. Moore had devised the pedagogical method now known as the "Moore Method," a process akin to the "heuristic" method described in the essay by Hadamard. There is evidence that Moore was skeptical of admitting into his course in topology a student who evinced interest in actuarial mathematics! Moore felt that the two disciplines were incompatible but he relented and had no reason ever to regret his decision. Indeed Wilder was Moore's first doctoral student at Texas, one of a long list of successful mathematicians.

Wilder got his PhD under Moore in 1923, stayed a year at Austin, then two years at Ohio State. There he had to cope with the requirement of taking an oath of allegiance including a pledge to uphold the laws of Ohio, which at that time included "prohibition laws." After two years at Ohio State, he went to the University of Michigan where he remained until statutory retirement in 1967. He then spent his retirement years at the University of California at Santa Barbara.

His creativity in a broad range of mathematical disciplines brought him wide recognition in the mathematical world—a Guggenheim fellowship, American Mathematical Society Colloquium lecturer, President of both the Mathematical Association of America as well as the American Mathematical Society. He was one of the invited speakers at the International Congress of Mathematicians in 1950 where he gave the address on the Cultural Basis of Mathematics.

The Wilders had four children, one of whom, Beth Dillingham, became an anthropologist who influenced Wilder in his search for a cultural basis of mathematics.

The Wilders were praised for their warm and gracious qualities. This humanism suffused their entire lives—lives filled with concern for the welfare of others. This attitude was especially expressed in his relations with students. He lived a fruitful life being granted more than the four score years allotted by the scriptures and died in 1982.

Editor's Preface

This essay, delivered at the International Congress of Mathematicians in Cambridge, Massachusetts in 1950, was the beginning of Wilder's continuing interest in the subject of mathematics as a cultural system. His ongoing interests culminated in a book of the above title published in 1974.

It is not easy to define what is meant by a "culture." Anthropologists are not in agreement as to the meaning. This much, however, can be said: there are patterns of behavior exhibited by different social groups and the behavior, often unique to that group, is part of its culture. As an example, a marriage ceremony is common to many different social groups but its form and significance comprise a part of the culture of the group.

In the case of the world of mathematicians as a group, the members are part of the wider world to which they belong. We must stress that Wilder is not referring to the type of mathematics germane to different social groups but to mathematics and mathematicians as a culture in themselves. They function according to patterns of behavior identified by anthropologists as characteristic of a "culture." This response is, according to Wilder, surprisingly uniform. It goes without saying that the wider social culture in which a mathematician lives inevitably has an influence on this response.

Nor can we separate cultural responses from the biological and psychological make up of humans. It would appear plausible to postulate an "inquisitiveness gene" which results in an inborn tendency of a small segment of society to want to examine why things are as they are. In addition, there is a "creative gene" which propels some to artistic creation or scientific invention. In appropriate circumstances, these find expression in mathematical activity. On the other hand, why some societies develop mathematics to a high degree while others do not, remains a mystery. Why these flourish in some societies and then decline is another mystery.

Wilder is more concerned however, with the anthropological behavior within the mathematical culture. What role for example, does fashion play? What impetus does competition provide? What influence does status exert?

Wilder has opened up a line of inquiry which has, in recent years, evoked considerable attention. His eminence as a mathematician has given the study of social, philosophical and psychological discourse about mathematics a "respectability" which it has not always had. These inquiries continue to flourish.

I presume that it is not inappropriate, on the occasion of an International Mathematical Congress which comes at the half-century mark, to devote a little time to a consideration of mathematics as a whole. The addresses and papers to be given in the various conferences and sectional meetings will in general be concerned with special fields or branches of mathematics. It is the aim of the present remarks to get outside mathematics, as it were, in the hope of attaining a new perspective. Mathematics has been studied extensively from the abstract philosophical viewpoint, and some benefits have accrued to mathematics from such studies—although generally the working mathematician is inclined to look upon philosophical speculation with suspicion. A growing number of mathematicians have been devoting thought to the Foundations of Mathematics, many of them mathematicians whose contributions to mathematics have won them respect. The varying degrees of dogmatism with which some of these have come to regard their theories, as well as the sometimes acrimonious debates which have occurred between holders of conflicting theories, makes one wonder if there is not some vantage point from which one can view such matters more dispassionately.

It has become commonplace today to say that mankind is in its present "deplorable" state because it has devoted so much of its energy to technical skills and so little to the study of man itself. Early in his civilized career, man studied astronomy and the other physical sciences, along with the mathematics that these subjects suggested; but in regard to such subjects as anatomy, for example, it was not easy for him to be objective. Man himself, it seemed, should be considered untouchable so far as his private person was concerned. It is virtually only within our own era that the study of the even more personal subjects, such as psychology, has become moderately respectable! But in the study of the behavior of man *en masse,* we have made little progress. This is evidently due to a variety of reasons such as (1) inability to distinguish between group behavior and individual behavior, and (2) the fact that although the average person may grudgingly give in to being cut open by a surgeon, or analyzed by a psychiatrist, those group institutions which determine his system of values, such as nation, church, clubs, etc., are still considered untouchable.

Fortunately, just as the body of the executed criminal ultimately became available to the anatomist, so the "primitive" tribes of Australia, the Pacific Islands, Africa, and the United States, became available to the anthropologist. Using methods that have now become so impersonal and objective as to merit its being classed among the natural sciences rather than with such social studies as history, anthropology has made great advances within the past 50 years in the study of the group behavior of mankind. Its development of the culture concept and investigation of cultural forces will, perhaps, rank among the greatest achievements of the human mind, and despite opposition, application of the concept has made strides in recent years. Not only are psychologists, psychiatrists, and sociologists applying it, but governments that seek to extend their control over alien peoples have recognized it. Manifold human suffering has resulted from ignorance of the concept, both in the treatment of colonial peoples, and in the handling of the American Indian, for example.

Now I am not going to offer the culture concept as an antidote for all the ills that beset mathematics. But I do believe that only by recognition of the cultural basis of mathematics will a better understanding of its nature be achieved; moreover, light can be thrown on various problems, particularly those of the Foundations of Mathematics. I don't mean that it can solve these problems, but that it can point the *way* to solutions as well as show the *kinds* of solutions that may be expected. In addition, many things that we have believed, and attributed to some kind of vague "intuition," acquire a real validity on the cultural basis.

For the sake of completeness, I shall begin with a rough explanation of the concept. (For a more adequate exposition, see [10; Chap. 7] and [18]. Obviously it has nothing to do with culture spelled with a "K," or with degrees from the best universities or inclusion in the "best" social circles. A culture is the collection of customs, rituals, beliefs, tools, mores, etc., which we may call *cultural elements,* possessed by a group of people, such as a primitive tribe or the people of North America. Generally it is not a fixed thing but changing with the course of time, forming what can be called a "culture stream." It is handed down from one generation to another, constituting a seemingly living body of tradition often more dictatorial in its hold than Hitler was over Nazi Germany; in some primitive tribes virtually every act, even such ordinary ones as eating and dressing, are governed by ritual. Many anthropologists have thought of a culture as a super-organic entity, having laws of development all its own, and most anthropologists seem in practice to treat a culture as a thing in itself, without necessarily referring (except for certain purposes) to the group or individuals possessing it.

We "civilized" people rarely think of how much we are dominated by our cultures—we take so much of our behavior as "natural." But if you were to propose to the average American male that he should wear earrings, you might, as you picked yourself off the ground, reflect on the reason for the blow that you have just sustained. Was it because he decided at some previous date that every time someone suggested wearing earrings to him he would respond with a punch to the nose? Of course not. It was decided for him and imposed on him by the American culture, so that what he did was, he would say, the "natural thing to do." However, there are societies such as Navajo, Pueblo, and certain Amazon tribes, for instance, in which the wearing of earrings by the males is the "natural thing to do." What we call "human nature" is virtually nothing but a collection of such culture traits. What is "human nature" for a Navajo is distinctly different from what is "human nature" for a Hottentot.

As mathematicians, we share a certain portion of our cultures which is called "mathematical." We are influenced by it, and in turn we influence it. As individuals we assimilate parts of it, our contacts with it being through teachers, journals, books, meetings such as this, and our colleagues. We contribute to its growth the results of our individual syntheses of the portions that we have assimilated .

Now to look at mathematics as a cultural element is not new. Anthropologists have done so, but as their knowledge of mathematics is generally very limited, their reactions have ordinarily consisted of scattered remarks concerning the types of arithmetic found in primitive cultures. An exception is an article [17] which appeared about three years ago, by the anthropologist L. A. White, entitled *The locus of mathematical reality,* which was inspired by the seemingly conflicting notions of the nature of mathematics as expressed by various mathematicians and philosophers. Thus, there is the belief expressed by G. H. Hardy [8; pp. 63–64] that "mathematical reality lies outside us, and that our function is to discover or *observe* it, and that the theorems which we prove, and which we describe grandiloquently as our creations are simply our notes of our observations." On the other hand there is the point of view expressed by P. W. Bridgman [3; p. 60] that "it is the merest truism, evident at once to unsophisticated observation, that mathematics is a human invention." Although these statements seem irreconcilable, such is not the case when they are suitably interpreted. For insofar as our mathematics is a part of our culture, it is, as Hardy says, "outside us." And insofar as a culture cannot exist except as the product of human minds, mathematics is, as Bridgman states, a "human invention."

As a body of knowledge, mathematics is not something I know, you know, or any individual knows: It is a part of our culture, our *collective* possession. We may even forget, with the passing of time, some of our own individual contributions to it, but these may remain, despite our forgetfulness, in the culture stream. As in the case of many other cultural elements, we are taught mathematics from the time when we are able to speak, and from the first we are impressed with what we call its "absolute truth." It comes to have the same significance and type of reality, perhaps, as their system of gods and rituals has for a primitive people. Such would seem to be the case of Hermite, for example, who according to Hadamard [7; p. xii] said, "We are rather servants than masters in Mathematics;" and who said [6; p. 449] in a letter to Königsberger,"—these notions of analysis have their existence apart from us,—they constitute a whole of which only a part is revealed to us, incontestably although mysteriously associated with that other totality of things which we perceive by way of the senses." Evidently Hermite sensed the impelling influence of the culture stream to which he contributed so much!

In his famous work *Der Untergang des Abendlandes* [15], O. Spengler discussed at considerable length the nature of mathematics and its importance in his organic theory of cultures. And under the influence of this work, C.J. Keyser published [9] some views concerning *Mathematics as a Culture Clue,* constituting an exposition and defense of the thesis that "The type of mathematics found in any major Culture is a clue, or key, to the distinctive character of the Culture taken as a whole." Insofar as mathematics is a part of and is influenced by the culture in which it is found, one may expect to find some sort of relationship between the two. As to how good a "key" it furnishes to a culture, however, I shall express no opinion; this is really a question for an anthropologist to answer. Since the culture dominates its elements, and in particular its mathematics, it would appear that for mathematicians it would be more fruitful to study the relationship from this point of view.

Let us look for a few minutes at the history of mathematics. I confess I know very little about it, since I am not a historian. I should think, however, that in writing a history of mathematics the historian would be constantly faced with the question of *what sort of material to include.* In order to make a clearer case, let us suppose that a hypothetical person, *A,* sets out to write a *complete* history, desiring to include all available material on the "history of mathematics." Obviously, he will have to accept some material and reject other material. It seems clear that his criterion for choice must be based on knowledge of what constitutes mathematics! If by this we mean a *definition* of mathematics, of course his task is hopeless. Many definitions have been given, but none has been chosen; judging by their number, it used to be expected of every self-respecting mathematician that he would leave a definition of mathematics to posterity! Consequently our hypothetical mathematician *A* will be guided, I imagine, by what is *called* "mathematics" in his culture, both in existing (previously written) histories and in works *called* "mathematical," as well as by what sort of thing people who are *called* "mathematicians" publish. He will, then, recognize what we have already stated, that mathematics is a certain part of his culture, and will be guided thereby.

For example, suppose *A* were a Chinese historian living about the year 1200 (500 or 1500 would do as well). He would include a great deal about computing with numbers and solving equations; but there wouldn't be any geometry as the Greek understood it in his history, simply because it had never been integrated with the mathematics of his culture. On the other hand, if *A* were a Greek of 200 A.D., his history of mathematics would be replete with geometry, but there would be little of algebra or even of computing with numbers as the Chinese practiced it. But if *A* were one of our contemporaries, he would include both geometry and algebra because both are part of what we call mathematics. I wonder what he would do about logic, however?

Here is a subject which, despite the dependence of the Greeks on logical deduction, and despite the fact that mathematicians, such as Leibniz and Pascal, have devoted considerable time to it on its own merits, has been given very little space in histories of mathematics. As an experiment, I looked in two histories that have been popular in this country; Ball's [1] and Cajori's [5], both written shortly before 1900. In the index of Ball's first edition (1888) there is no mention of "logic"; but in the fourth edition (1908) "symbolic and mathematical logic" is mentioned with a single citation, which proved to be a reference to an incidental remark about George Boole to the effect that he "was one of the creators of symbolic or mathematical logic." Thus symbolic logic barely squeezed under the line because Boole was a mathematician! The index of Cajori's first edition (1893) contains four citations under "logic," all referring to incidental remarks in the text. None of these citations is repeated in the second edition (1919), whose index has only three citations under "logic" (two of which also constitute the sole citations under "symbolic logic"), again referring only to brief remarks in the text. Inspection of the text, however, reveals nearly four pages (407–410) of material under the title "Mathematical logic," although there is no citation to this subject in the index nor is it cited under "logic" or "symbolic logic." (It is as though the subject had, by 1919, achieved enough importance for inclusion as textual material in a history of mathematics although not for citation in the index!)

I doubt if a like situation could prevail in a history of mathematics which covers the past 50 years! The only such history that covers this period, that I am acquainted with, is Bell's *Development of Mathematics* [2]. Turning to the index of this book, I found so many citations to "logic" that I did not care to count them. In particular, Bell devotes at least 25 pages to the development of what he calls "mathematical logic." Can there be any possible doubt that this subject, not considered part of mathematics in our culture in 1900, despite the pioneering work of Peano and his colleagues, is now in such "good standing" that any impartial definition of mathematics must be broad enough to include it?

Despite the tendency to approach the history of mathematics from the biographical standpoint, there has usually existed some awareness of the impact of cultural forces. For example, in commencing his chapter on Renaissance mathematics, Ball points out the influence of the introduction of the printing press. In the latest histories, namely the work of Bell already cited, and Struik's excellent little two volume work [16], the evidence is especially strong. For example, in his introduction, Struik expresses regret that space limitations prevented sufficient "reference to the general cultural and sociological atmosphere in which the mathematics of a period matured—or was stifled." And he goes on to say "Mathematics has been influenced by agriculture, commerce and manufacture, by warfare, engineering and philosophy, by physics and by astronomy. The influence of hydrodynamics on function theory, of Kantianism and of surveying on geometry, of electromagnetism on differential equations, of Cartesianism on mechanics, and of scholasticism on the calculus could only be indicated [in his book] ;—yet an understanding of the course and content of mathematics can be reached only if all these determining factors are taken into consideration." In his third chapter Struik gives a revealing account of the rise of Hellenistic mathematics, relating it to the cultural conditions then prevailing. I hope that future histories of mathematics will similarly give more attention to mathematics as a cultural element, placing greater emphasis on its relations to the cultures in which it is imbedded.

In discussing the general culture concept, I did not mention the two major processes of cultural change, *evolution* and *diffusion*. By diffusion is meant the transmission of a cultural trait from one culture to another, as a result of some kind of contact between groups of people; for example, the diffusion of French language and customs into the Anglo-Saxon culture following the Norman conquest. As to how much of what we call cultural progress is due to evolution and how much to diffusion, or to a combination of both, is usually difficult to determine, since the two processes tend so much to merge. Consider, for example, the counting process. This is what the anthropologist calls a universal trait—what I would prefer to call, in talking to mathematicians, a cultural invariant—it is found in every culture in at least a rudimentary form. The "base" may be 10, 12, 20, 25, 60—all of these are common, and are evidently determined by other (variable) culture elements—but the counting process in its essence, as the Intuitionist speaks of it, is invariant. If we consider more advanced cultures, the notion of a zero element sometimes appears. As pointed out by the anthropologist A. L. Kroeber, who in his *Anthropology* calls it a "milestone of civilization," a symbol for zero evolved in the cultures of at least three peoples; the Neo-Babylonian (who used a sexagesimal system), the Mayan (who used a vigesimal system), and the Hindu (from whom our decimal system is derived) [10; pp. 468–472]. Attempts by the extreme "diffusionists" to relate these have not yet been successful, and until they are, we can surmise that the concept of zero might ultimately evolve in any culture.

The Chinese-Japanese mathematics is of interest here. Evidently, as pointed out by Mikami [13] and others, the Chinese borrowed the zero concept from the Hindus, with whom they established contact at least as early as the first century, A.D. Here we have an example of its introduction by diffusion, but without such contacts, the zero would probably have evolved in Chinese mathematics, especially since calculators of the rod type were employed. The Chinese mathematics is also interesting from another standpoint in that its development seems to have been so much due to evolution within its own culture and so little affected by diffusion. Through the centuries it developed along slender arithmetic and algebraic lines, with no hint of geometry as the Greeks developed it. Those who feel that without the benefit of diffusion a culture will eventually stagnate find some evidence perhaps in the delight with which Japanese mathematicians of the 17th and 18th centuries, to whom the Chinese mathematics had come by the diffusion process, solved equations of degrees as high as 3000 or 4000. One is tempted to speculate what might have happened if the Babylonian zero and method of position had been integrated with the *Greek* mathematics—would it have meant that Greek mathematics might have taken an algebraic turn? Its introduction into the Chinese mathematics certainly was not productive, other than in the slight impetus it gave an already computational tendency.

That the Greek mathematics was a natural concomitant of the other elements in Greek culture, as well as a natural result of the evolution and diffusion processes that had produced this culture in the Asia Minor area, has been generally recognized. Not only was the Greek culture conducive to the type of mathematics that evolved in Greece, but it is probable that it resisted integration with the Babylonian method of enumeration. For if the latter became known to certain Greek scholars, as some seem to think, its value could not have been apparent to the Greeks.

We are familiar with the manner in which the Hindu-Arabic mathematical cultures diffused via Africa to Spain and then into the Western European cultures. What had become stagnant came to life—analytic geometry appeared, calculus—and the flood was on. The mathematical cultural development of these times would be a fascinating study, and awaits the cultural historian who will undertake it. The easy explanation that a number of "supermen" suddenly appeared on the scene has been abandoned by virtually all anthropologists. A *necessary* condition for the emergence of the "great man" is the presence of suitable cultural environment, including opportunity, incentive, and materials. Who can doubt that potentially great algebraists lived in Greece? But in Greece, although the opportunity and incentive may have been present, the cultural materials did not contain the proper symbolic apparatus. The anthropologist Ralph Linton remarked [12; p. 319] "The mathematical genius can only carry on from the point which mathematical knowledge within his culture has already reached. Thus if Einstein had been born into a primitive tribe which was unable to count beyond three, life-long application to mathematics probably would not have carried him beyond the development of a decimal system based on fingers and toes." Furthermore, the evidence points strongly to the *sufficiency of* the conditions stated: That is, suitable cultural environment is sufficient for the emergence of the great man. If your philosophy depends on the assumption of free will, you can probably adjust to this. For certainly your will is no freer than the opportunity to express it; you may will a trip to the moon this evening, but you won't make it. There may be potentially great blancophrenologists sitting right in this room; but if so they are destined to go unnoticed and undeveloped because blancophrenology is not yet one of our cultural elements.

Spengler states it this way [15tr; vol. II, p. 507]: "We have not the freedom to reach to this or to that, but the freedom to do the necessary or to do nothing. And a task that historic necessity has set *will* be accomplished with the individual or against him." As a matter of fact, when a culture or cultural element has developed to the point where it is ready for an important innovation, the latter is likely to emerge in more than one spot. A classical example is that of the theory of biological evolution, which had been anticipated by Spencer and, had it not been announced by Darwin, was ready to be announced by Wallace and soon thereafter by others. And as in this case, so in most other cases,—and you can recall many such in mathematics—one can after the fact usually go back and map out the evolution of the theory by its traces in the writings of men in the field.

Why are so many giving their lives to mathematics today; why have the past 50 years been so productive mathematically? The mathematical groundwork laid by our predecessors, the universities, societies, foundations, libraries, etc., have furnished unusual opportunity, incentive, and cultural material. In addition, the processes of evolution and diffusion have greatly accelerated. Of the two, the latter seems to have played the greater role in the recent activity. For during the past 50 years there has been an exceptional amount of fusion of different branches of mathematics, as you well know. A most unusual cultural factor affecting the development of mathematics has been the emigration of eminent mathematicians from Germany, Poland, and other countries to the United States during the past 30 years. Men whose interests had been in different branches of mathematics were thrown together and discovered how to merge these branches to their mutual benefit, and frequently new branches grew out of such meetings. The cultural history of mathematics during the past 50 years, taken in conjunction with that of mathematics in ancient Greece, China, and Western Europe, furnishes convincing evidence that *no branch of mathematics can pursue its course in isolation indefinitely, without ultimately reaching a static condition.*

Of the *instruments* for diffusion in mathematics, none is more important, probably, than the journals. Without sufficient outlet for the results of research, and proper distribution of the same, the progress of mathematics will be severely hampered. And any move that retards international contacts through the medium of journals, such as restriction to languages not widely read, is distinctly an anti-mathematical act. For it has become a truism that today mathematics is international.

This brings us to a consideration of symbols. For the so-called "international character" of mathematics is due in large measure to the standardization of symbols that it has achieved, thereby stimulating diffusion. Without a symbolic apparatus to convey our ideas to one another, and to pass on our results to future generations, there wouldn't be any such thing as mathematics—indeed, there would be essentially no culture at all, since, with the possible exception of a few simple tools, culture is based on the use of symbols. A good case can be made for the thesis that man is to be distinguished from other animals by the way in which he uses symbols [18; II]. Man possesses what we might call *symbolic initiative;* that is, he assigns symbols to stand for objects or ideas, sets up relationships between them, and operates with them as though they were physical objects. So far as we can tell, no other animal has this faculty, although many animals do exhibit what we might call *symbolic reflex* behavior. Thus, a dog can be taught to lie down at the command "Lie down," and of course to Pavlov's dogs, the bells signified food. In a recent issue of a certain popular magazine a psychologist is portrayed teaching pigeons to procure food by pressing certain combinations of colored buttons. All of these are examples of symbolic. reflex behavior—the animals do not create the symbols.

As an aspect of our culture that depends so exclusively on symbols, as well as the investigation of certain relationships between them, mathematics is probably the furthest from comprehension by the non-human animal. However, much of our mathematical behavior that was originally of the *symbolic initiative* type drops to the *symbolic reflex* level. This is apparently a kind of labor-saving device set up by our neural systems. It is largely due to this, I believe, that a considerable amount of what passes for "good" teaching in mathematics is of the symbolic reflex type, involving no use of symbolic initiative. I refer of course to the drill type of teaching which may enable stupid John to get a required credit in mathematics but bores the creative minded William to the extent that he comes to loathe the subject! What essential difference is there between teaching a human animal to take the square root of 2 and teaching a pigeon to punch certain combinations of colored buttons? Undoubtedly the *symbolic reflex* type of teaching is justified when the pupil is very young—closer to the so-called "animal" stage of his development, as we say. But as he approaches maturity, more emphasis should be placed on his symbolic initiative. I am reminded here of a certain mathematician who seems to have an uncanny skill for discovering mathematical talent among the undergraduates at his university. But there is nothing mysterious about this; he simply encourages them to use their symbolic initiative. Let me recall parenthetically here what I said about the perennial presence of potential "great men"; there is no reason to believe that this teacher's success is due to a preference for his university by the possessors of mathematical talent, for they usually have no intention of becoming mathematicians when they matriculate. It moves one to wonder how many potentially great mathematicians are being constantly lost to mathematics because of "symbolic reflex" types of teaching.

I want to come now to a consideration of the Foundations of Mathematics. We have witnessed, during the past 50 years, what we might call the most thorough soul-searching in the history of mathematics. By 1900, the Burali-Forti contradiction had been found and the Russell and other antinomies were soon to appear. The sequel is well known: Best known are the attempt of Russell and Whitehead in their monumental *Principia Mathematica* to show that mathematics can be founded, in a manner free of contradiction, on the symbolically expressed principles and methods of what were at the time considered universally valid logical concepts; the formulation, chiefly at the hands of Brouwer and his collaborators, of the tenets of Intuitionism, which although furnishing a theory evidently free of contradiction, introduces a highly complicated set theory and a mathematics radically restricted as compared with the mathematics developed during the 19th century; and the formalization of mathematics by Hilbert and his collaborators, together with the development of a metamathematical proof theory which it was hoped would lead to proofs of freedom from contradiction for a satisfactory portion, at least, of the classical mathematics. None of these "foundations" has met with complete success. Russell and Whitehead's theory of types had to be bolstered with an axiom which they had to admit, in the second edition of *Principia Mathematica,* has only pragmatic justification, and subsequent attempts by Chwistek, Wittgenstein, and Ramsey to eliminate or modify the use of this axiom generally led to new objections. The restricted mathematics known as Intuitionism has won only a small following, although some of its methods, such as those of a finite constructive character, seem to parallel the methods underlying the treatment of formal systems in symbolic logic, and some of its tenets, especially regarding constructive existence proofs, have found considerable favor. The possibility of carrying out the Hilbert program seems highly doubtful, in view of the investigations of Gödel and others.

Now the cultural point of view is not advanced as a substitute for such theories. In my title I have used the word "basis" instead of "foundations" in order to emphasize this point. But it seems probable that the recognition of the cultural basis of mathematics would clear the air in foundation theories of most of the mystical and vague philosophical arguments which are offered in their defense, as well as furnish a guide and motive for further research. The points of view underlying various attempts at Foundations of Mathematics are often hard to comprehend. In most cases it would seem that the proponents have decided in their own minds just what mathematics is, and that all they have to do is formulate it accordingly—overlooking entirely the fact that because of its cultural basis, mathematics as they know it may be not at all what it will be a century hence. If the thought underlying their endeavors is that they will succeed in trapping the elusive beast and confining it within bounds which it will never break, they are exceedingly optimistic. If the culture concept tells us anything, it should teach us that the first rule for setting up any Foundation theory is that it should only attempt to encompass specific portions of the field as it is known in our culture. At most, a Foundation theory should be considered as a kind of constitution with provision for future amendments. And in view of the situation as regards such principles as the choice axiom, for instance, it looks at present as though no such constitution could be adopted by a unanimous vote!

I mentioned "mysticism and vague philosophical arguments" and their elimination on the cultural basis. Consider, for example, the insistence of Intuitionism that all mathematics should be founded on the natural numbers or the counting process, and that the latter are "intuitively given." There are plausible arguments to support the thesis that the natural numbers should form the starting point for mathematics but it is hard to understand just what "intuitively given" means, or why the classical conception of the continuum, which the Intuitionist refuses to accept, should not be considered as "intuitively given." It makes one feel that the Intuitionist has taken Kronecker's much-quoted dictum that "The integers were made by God, but all else is the work of man" and substituted "Intuition" for "God." However, if he would substitute for this vague psychological notion of "intuition" the viewpoint that inasmuch as the counting process is a cultural invariant, it follows that the natural numbers form for every culture the most basic part of what has been universally called "mathematics," and should therefore serve as the starting point for every Foundations theory; then I think he would have a much sounder argument. I confess that I have not studied the question as to whether he can find further cultural support to meet all the objections of opponents of Intuitionism. It would seem, however, that he would have to drop his insistence that in construction of sets (to quote Brouwer [4; p. 86]) "neither ordinary language nor any symbolic language can have any other role than that of serving as a non-mathematical auxiliary," since no cultural trait on the abstract level of mathematics can be constructed other than by the use of symbols. Furthermore, and this is a serious objection, it appears to ignore *the influence that our language habits have on our modes of thought*.

Or consider the thesis that all mathematics is derivable from what some seem to regard as primitive or universal logical principles and methods. Whence comes this "primitive" or "universal" character? If by these terms it is meant to imply that these principles have a culturally invariant basis like that of the counting process, then it should be pointed out that cultures exist in which they do not have any validity, even in their qualitative non-symbolic form. For example, in cultures which contain magical elements (and such elements form an extremely important part of some primitive cultures), the law of contradiction usually fails. Moreover, the belief that our forms of thought are culturally invariant is no longer held. As eminent a philosopher as John Stuart Mill stated, [14; p. 11], "The principles and rules of grammar are the means by which the forms of language are made to correspond with the universal forms of thought." If Mill had been acquainted with other than the Indo-European language group, he could not have made such an error. The Trobriand Islanders, for example, lack a cause-and-effect pattern of thought; their language embodies no mechanism for expressing a relationship between events. As Malinowski pointed out [11; p. 360], these people have no conception of one event leading up to another, and chronological sequence is unimportant. (Followers of Kant should note that they can count, however.) But I hardly need to belabor the point. As Lukasciewicz and others have observed, not even Aristotle gave to the law of the excluded middle the homage that later logicians paid it! All I want to do in this connection is to indicate that on the cultural basis we find affirmation of what is already finding universal acceptance among mathematical logicians, I believe; namely, that the significance and validity of such material as that in *Principia Mathematica* is only the same as that of other purely formal systems.

It is probably fair to say that the Foundations of Mathematics as conceived and currently investigated by the mathematical logicians finds greatest support on the cultural basis. For inasmuch as there can exist, and have existed, different cultures, different forms of thought, and hence different mathematics, it seems impossible to consider mathematics, as I have already indicated, other than man-made and having no more of the character of necessity or truth than other cultural traits. Problems of mathematical existence, for example, can never be settled by appeal to any mathematical dogma. Indeed, they have no validity except as related to special foundations theories. The question as to the existence of choice sets, for instance, is not the same for an Intuitionist as for a Formalist. The Intuitionist can justifiably assert that "there is no such problem as the continuum problem" provided he adds the words "for an Intuitionist"—otherwise he is talking nonsense. *Because of its cultural basis, there is no such thing as the absolute in mathematics; there is only the relative.*

But we must not be misled by these considerations and jump to the conclusion that what constitutes mathematics in our culture is purely arbitrary; that, for instance, it can be defined as the "science of p implies q," or the science of axiomatic systems. Although the individual person in the cultural group may have some degree of variability allowed him, he is at the same time subject to the dominance of his culture. The individual mathematician can play with postulational systems as he will, but unless and until they are related to the existing state of mathematics in his culture they will only be regarded as idiosyncrasies. Similar ties, not so obvious however, join mathematics to other cultural elements. And these bonds, together with those that tie each and every one of us to our separate mathematical interests, cannot be ignored even if we will to do so. They may exert their influence quite openly, as in the case of those mathematicians who have recently been devoting their time to high speed computers, or to developing other new and unforeseen mathematics induced by the recent wartime demands of our culture. Or their influence may be hidden, as in the case of certain mathematical habits which were culturally induced and have reached the symbolic reflex level in our reactions. Thus, although the postulational method may turn out to be the most generally accepted mode of founding a theory, it must be used with discretion; otherwise the theories produced will not be mathematics in the sense that they will be a part of the mathematical component of our culture.

But it is time that I closed these remarks. It would be interesting to study evidence in mathematics of *styles* and of *cultural patterns;* these would probably be interesting subjects of investigation for either the mathematician or the anthropologist, and could conceivably throw some light on the probable future course of the field. I shall have to pass on, however, to a brief conclusion:

In man's various cultures are found certain elements which are called *mathematical*. In the earlier days of civilization, they varied greatly from one culture to another so much so that what was called "mathematics" in one culture would hardly be recognized as such in certain others. With the increase in diffusion due, first, to exploration and invention, and, secondly, to the increase in the use of suitable symbols and their subsequent standardization and dissemination in journals, the mathematical elements of the most advanced cultures gradually merged until, except for minor cultural differences like the emphasis on geometry in Italy, or on function theory in France, there has resulted essentially one element, common to all civilized cultures, known as mathematics. This is not a fixed entity, however, but is subject to constant change. Not all of the change represents accretion of new material; some of it is a shedding of material no longer, due to influential cultural variations, considered mathematics. Some so-called "borderline" work, for example, it is difficult to place either in mathematics or outside mathematics.

From the extension of the notion of number to the transfinite, during the latter half of the 19th century, there evolved certain contradictions around the turn of the century, and as a consequence the study of Foundations questions, accompanied by a great development of mathematical logic, has increased during the last 50 years. Insofar as the search for satisfactory Foundation theories aims at any absolute criterion for truth in mathematics or fixation of mathematical method, it appears doomed to failure, since recognition of the cultural basis of mathematics compels the realization of its variable and growing character. Like other culture traits, however, mathematics is not a thoroughly arbitrary construction of the individual mathematician, since the latter is restricted in his seemingly free creations by the state of . mathematics and its directions of growth during his lifetime, it being the latter that determines what is considered "important" at the given time.

In turn, the state and directions of growth of mathematics are determined by the general complex of cultural forces both within and without mathematics. Conspicuous among the forces operating from without during the past 50 years have been the crises through which the cultures chiefly concerned have been passing; these have brought about a large exodus of mathematicians from Western Europe to the United States, thereby setting up new contacts with resulting diffusion and interaction of mathematical ideas, as well as in the institution of new directions or acceleration of directions already under way, such as in certain branches of applied mathematics.

What the next 50 years will bring, I am not competent to predict. In his *Decline of the West,* Spengler concluded [15tr; pp. 89–90] that in the notion of *group,* Western "mathematic" had achieved its "last and conclusive creation," and he closed his second chapter, entitled "The meaning of numbers," with the words: "—the time of the *great* mathematicians is past. Our tasks today are those of preserving, rounding off, refining, selection—in place of big dynamic creation, the same clever detail-work which characterized the Alexandrian mathematics of late Hellenism." This was published in 1918—32 years ago—and I leave it to your judgment whether he was right or not. It seems unlikely that the threatened division into two opposing camps of those nations in which mathematical activity is chiefly centered at present will be of long enough duration to set up two distinct mathematical cultures—although in other fields, such as botany, such a division appears to be under way. Nevertheless, as individual mathematicians we are just as susceptible to cultural forces as are botanists, economists, or farmers, and long separation in differing cultures can result in variations of personality that cannot fail to be reflected in our mathematical behavior. Let us hope that at the turn of the century 50 years hence, mathematics will be as active and unique a cultural force as it is now, with that free dissemination of ideas which is the chief determinant of growth and vitality.

I am indebted to my colleague, Professor L. A. White, of the Anthropology Department, University of Michigan, and to Betty Ann Dillingham, for reading this paper in manuscript and offering most helpful criticism and advice. However, responsibility for errors and opinions expressed herein is entirely my own, of course.

Bibliography

1. W. W. R. Ball, *A short account of the history of mathematics,* London, Macmillan, 1888; 4th ed., 1908.
2. E. T. Bell, *The development of mathematics,* New York, McGraw-Hill, 2d ed., 1945.
3. P. W. Bridgman, *The logic of modern physics,* New York, Macmillan, 1927.
4. L. E. J. Brouwer, Intuitionism and formalism, *Bull. Amer. Math. Soc.* 20 (1913–1914) 81–96, (tr. by A. Dresden).
5. F. Cajori, *A history of mathematics,* New York, Macmillan, 1893; 2d ed., 1919.

6. A. Dresden, Some philosophical aspects of mathematics, *Bull. Amer. Math. Soc.* 34 (1928) 438–452.
7. J. Hadamard, *The psychology of invention in the mathematical field,* Princeton, Princeton University Press, 1945.
8. G. H. Hardy, *A mathematician's apology,* Cambridge, England, Cambridge University Press, 1941.
9. C. J. Keyser, Mathematics as a culture clue, *Scripta Mathematica* 1 (1932–1933) 185–203; reprinted in a volume of essays having same title, New York, *Scripta Mathematica,* 1947.
10. A. L. Kroeber, *Anthropology,* rev. ed, New York, Harcourt, Brace, 1948.
11. D. D. Lee, A primitive system of values, *Philosophy of Science* 7 (1940) 355–378.
12. R. Linton, *The study of man,* New York, Appleton-Century, 1936.
13. Y. Mikami, *The development of mathematics in China and Japan,* Leipzig, Drugulin, 1913.
14. J. S. Mill, *Inaugural address,* delivered to the University of St. Andrews, Feb. 1, 1867, Boston, Littell and Gay.
15. O. Spengler, *Der Untergang des Abendlandes,* München, C. H. Beck, vol. I, 1918, (2d ed., 1923), vol. II, 1922. [15tr] English translation of [15] by C. F. Atkinson, under the title *The decline of the West,* New York, Knopf, vol. I, 1926, vol. II, 1928.
16. D. J. Struik, *A concise history of mathematics,* 2 vols., New York, Dover, 1948.
17. L. A. White, The locus of mathematical reality, *Philosophy of Science* 14 (1947) 289–303; republished in somewhat altered form as Chapter 10 of [18].
18. *The science of culture,* New York, Farrar, Straus, 1949.

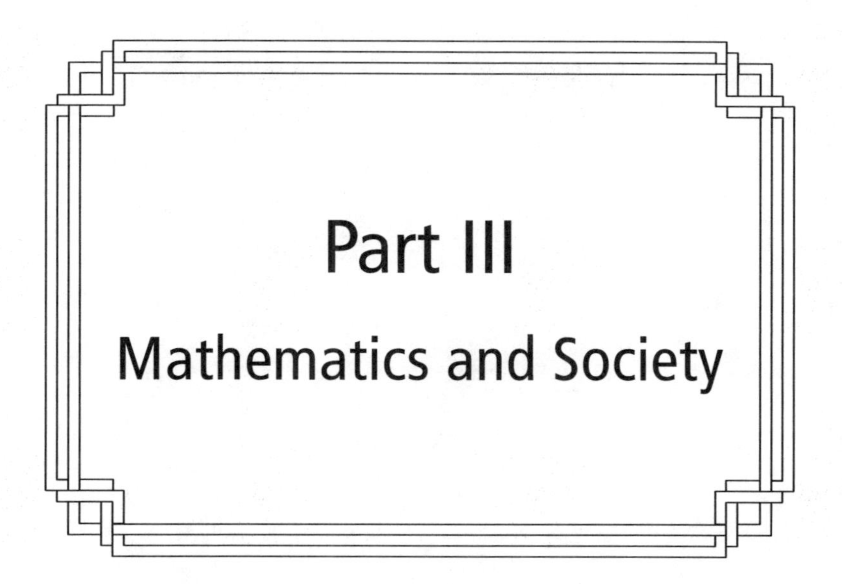

Part III

Mathematics and Society

Mathematics seems to endow one with something like a new sense. — Charles Darwin (1809–1882)

Presidential Address to the British Association

James Joseph Sylvester

J. J. Sylvester (1814–1897) is one of the most colorful representatives in the pantheon of mathematicians. He was born in London, the youngest of nine children born to Abraham Joseph and his wife. The oldest son had immigrated to America and had changed his surname to Sylvester and the English family later did likewise. James entered Cambridge University and although he attended lectures and sat for the tripos exam, in which he ranked second (second wrangler), he was not granted a degree as such, since the granting of a degree at Cambridge was, at that time, subject to the Universities Test. This "test" required oaths of Allegiance and Supremacy on the part of students and faculty. These oaths certified the candidate's willingness to worship and practice according to the communion of the Church of England, that is, the Anglican Church. Being Jewish, Sylvester could not take such oaths. This same requirement also impeded his ability later to secure a university appointment. The Test Act, as this was called, was repealed in 1871. It is interesting to note that he was awarded an M.A. degree by Trinity College, Dublin, although he had not attended classes there.

He had studied law and while practicing it, he met his collaborator to be, Arthur Cayley. At one time he also earned a living as an actuary. As professor of natural philosophy, he taught physics for three years at the University of London beginning in 1838. Tiring of physics, he applied for and received an appointment in mathematics at the University of Virginia but this, alas, was short-lived. There are apocryphal accounts as to why he left so precipitously. One of these stories points to an altercation with some students over the issue of slavery to which he was unalterably opposed.

He returned to England where he was appointed to a professorship at the Royal Military College in Woolwich, a post he held from 1855 to 1870. Forced to retire on account of age, Sylvester spent the next six years writing and commenting on poetry. In 1876 he was appointed to a professorship at Johns Hopkins in Baltimore. While there, he founded and became the first editor of the *American Journal of Mathematics.* He returned to England in 1883 when he was appointed to the Savilian chair of mathematics at Oxford.

He was endowed with a somewhat flamboyant personality and his love of language could be seen in his mathematical works. He left his mark on some of the terminology in algebra. He also wrote poetry and published a book on *The Laws of Verse.* Although his mathematical interests were wide-ranging (as were all his other interests), his main mathematical contributions were in the area of algebra, especially in matrix theory in which he collaborated with Cayley and where he left a legacy that is still the focus of some mathematical research. He also made significant contributions to the theory of partitions. Parenthetically it should be added that another of his claims to fame is that he was once tutor to Florence Nightingale.

He died in 1897.

Editor's Preface

This seemingly disjointed lecture is one Sylvester gave when he was appointed president of the British Association. It is typical of his writing, whether literary or mathematical. Endowed with an encyclopedic knowledge and a fluid mind, he could expostulate at will on a wide variety of subjects, often interjecting comments tangential to his main theme. Indeed, he was one of William James's predecessors in creating the style later called "stream of consciousness."

After a slow start, he begins by refuting a statement made by the noted biologist T. H. Huxley. The latter says that mathematics is simply a sequence of deductions from self-evident propositions and that it knows nothing of observation, experiment, induction and causation. (Huxley's father was a mathematics teacher). On the contrary says Sylvester, mathematics is constantly invoking new principles, new ideas, new methods which are not readily articulated but which have their source in the human mind. The extent to which observation contributes to new concepts and results is not easy to assess. That it plays a role in geometry is credible; that it plays a role in algebra and analysis appears less convincing but Sylvester is persuaded of its efficacy. Can one draw inference, or at least conjectures, simply by observation? According to Sylvester, the mathematicians Lagrange, Gauss, Riemann and others expressed their faith in the powers of observation. In many instances this may well be the case. A simple example for the non–mathematician will serve to illustrate the point. Looking at the sequence, 1, 1, 2, 3, 5, 8, 13, 21,..., one easily surmises what the law of formation might be. This phenomenon might well be the case for more complex examples.

In the realm of geometry, Sylvester says that Riemann, like his predecessor Gauss, rejected the Kantian doctrine that space and time are forms of intuition; they view these concepts as being possessed of objective reality. To bolster his view on the effectiveness of observation as an instrument of discovery, Sylvester cites the law of quadratic reciprocity and the existence of doubly periodic functions.

Sylvester decries what he calls the outmoded methods of teaching in Britain as contrasted with those on the continent. Pedagogy he says, should be more a one on one experience, a manner "unknown to the frozen formality of the present educational system." Followed in this manner, the study of mathematics "brings into harmonious action all the faculties of the mind and leads to higher states of conscious intellectual being." He boldly suggests that this intellectual exercise might account for the extraordinary longevity of the great practitioners of the mathematical art!

He proposes "honorably shelving" Euclid and replacing it in the curriculum with the more vital topics of contemporary algebra. This suggestion was made in recent decades by the noted French mathematician Jean Dieudonné. The latter's motives for doing so were not unlike those of Sylvester, i.e., an effort to revitalize mathematics in the schools.

This is an interesting statement from an unusually creative and fertile mind—a mind that failed to follow advice, given at the end of the article, to succeeding speakers, viz., "the importance of practicing brevity and condensation...."!

Ladies and Gentlemen,

A few days ago I noticed in a shop window the photograph of a Royal mother and child, which seemed to me a very beautiful group; on scanning it more closely, I discovered that the faces were ordinary, or, at all events, not much above the average, and that the charm arose entirely from the natural action and expression of the mother stooping over and kissing her child which she held in her lap; and I remarked to myself that the homeliest features would become beautiful when lit up by the rays of the soul—like the sun "gilding pale streams with heavenly alchemy." By analogy, the thought struck me that if a man would speak naturally and as he felt on any subject of his predilection, he might hope to awaken a sympathetic interest in the minds of his hearers; and, in illustration of this, I remembered witnessing how the writer of a well-known article in the *Quarterly Review* so magnetized his audience at the Royal Institution by his evident enthusiasm that, when the lecture was over and the applause had subsided, some ladies came up to me and implored me to tell them what they should do to get up the Talmud; for that was what the lecture had been about.

Now, as I believe that even Mathematics are not much more repugnant than the Talmud to the common apprehension of mankind, and I really love my subject, I shall not quite despair of rousing and retaining your attention for a short time if I proceed to read (as, for greater assurance against breaking down, I shall beg your permission to do) from the pages I hold in my hand.

It is not without a feeling of surprise and trepidation at my own temerity that I find myself in the position of one about to address this numerous and distinguished assembly. When informed that the Council of the British Association had it in contemplation to recommend me to the General Committee to fill the office of President of the Mathematical and Physical Section, the intimation was accompanied with the tranquilizing assurance that it would rest with myself to deliver or withhold an address as I might think fit, and that I should be only following in the footsteps of many of the most distinguished of my predecessors were I to resolve on the latter course.

Until the last few days I had made up my mind to avail myself of this option, by proceeding at once to the business before us without troubling you to listen to any address, swayed thereto partly by a consciousness of the very limited extent of my oratorical powers, partly by a disinclination, in the midst of various pressing private and official occupations, to undertake a kind of work new to one more used to thinking than to speaking (to making mathematics than to talking about them), and partly and more especially by a feeling of my inadequacy to satisfy the expectations that would be raised in the minds of those who had enjoyed the privilege of hearing or reading the allocution (which fills me with admiration and dismay) of my gifted predecessor, Dr Tyndall, a man in whom eloquence and philosophy seem to be inborn, whom Science and Poetry woo with an equal spell [2] and whose ideas have a faculty of arranging themselves in forms of order and beauty as spontaneously and unfailingly as those crystalline solutions from which, in a striking passage of his address, he drew so vivid and instructive an illustration.

From this lotus-eater's dream of fancied security and repose I was rudely awakened by receiving from the Editor of an old-established journal in this city a note containing a polite but peremptory request that I should, at my earliest convenience, favor him with a "copy of the address I proposed to deliver at the forthcoming Meeting." To this invitation, my first impulse was to respond very much in the same way as did the "Needy knife-grinder" of the *Antijacobin,* when summoned to recount the story of his wrongs to his republican sympathizer, "Story, God bless you, I have none to tell, Sir!" "Address, Mr Editor, I have none to deliver."

I have found, however, that increase of appetite still grows with what it feeds on, that those who were present at the opening of the Section last year, and enjoyed my friend Dr Tyndall's melodious utterances, would consider themselves somewhat ill-treated if they were sent away quite empty on the present occasion, and that, failing an address, the Members would feel very much like the guests at a wedding-breakfast where no one was willing or able to propose the health of the bride and bridegroom.

Yielding, therefore, to these considerations and to the advice of some officially connected with the Association, to whose opinions I feel bound to defer, and unwilling also to countenance by my example the too prevailing opinion that mathematical pursuits unfit a person for the discharge of the common duties of life and cut him off from the exercise of Man's highest prerogative, "discourse of reason and faculty of speech divine,"—rather, I say, than favor the notion that we Algebraists (who regard each other as the flower and salt of the earth) are a set of mere calculating-machines endowed with organs of locomotion, or, at best, a sort of poor visionary dumb creatures only capable of communicating by signs and symbols with the outer world, I have resolved to take heart of grace and to say a few words, which I hope to render, if not interesting, at least intelligible, on a subject to which the larger part of my life has been devoted.

The President of the Association, Prof. Stokes, is so eminent alike as a mathematician and physicist, and so distinguished for accuracy and extent of erudition and research, that I felt assured I might safely assume he would, in his Address to the Association at large, take an exhaustive survey, and render a complete account of the recent progress and present condition and prospects of Mathematical and Physical Science. This consideration narrowed very much and brought almost to a point the ground available for me to occupy in this Section; and as I cannot but be aware that it is as a cultivator of pure mathematics (the subject in which my own researches have chiefly, though by no means exclusively lain [3], that I have been placed in this Chair, I hope the Section will patiently bear with me in the observations I shall venture to make on the nature of that province of the human reason and its title to the esteem and veneration with which through countless ages it has been and, so long as Man respects the intellectual part of his nature, must ever continue to be regarded. [4]

It is said of a great party leader and orator in the House of Lords that, "when lately requested to make a speech at some religious or charitable (at all events a non-political) meeting, he declined to do so on the ground that he could not speak unless he saw an adversary before him—somebody to attack or reply to." In obedience to a somewhat similar combative instinct, I set to myself the task of considering certain recent utterances of a most distinguished member of this Association, one whom I no less respect for his honesty and public spirit than I admire for his genius and eloquence [5], but from whose opinions on a subject which he has not studied I feel constrained to differ. Goethe has said-

"Verständige Leute kannst du irren sehn
In Sachen, nämlich, die sie nicht verstehn."

Understanding people you may see erring in those things, to wit, which they do not understand.

I have no doubt that had my distinguished friend, the probable President-elect of the next Meeting of the Association, applied his uncommon powers of reasoning, induction, comparison, observation, and invention to the study of mathematical science, he would have become as great a mathematician as he is now a biologist; indeed he has given public evidence of his ability to grapple with the practical side of certain mathematical questions; but he has not made a study of mathematical science as such, and the eminence of his position and the weight justly attaching to his name render it only the more imperative that any assertions proceeding from such a quarter, which may appear to me erroneous, or so expressed as to be conducive to error, should not remain unchallenged or be passed over in silence.[6]

He says "mathematical training is almost purely deductive. The mathematician starts with a few simple propositions, the proof of which is so obvious that they are called self-evident, and the rest of his work consists of subtle deductions from them. The teaching of languages, at any rate as ordinarily practiced, is of the same general nature—authority and tradition furnish the data, and the mental operations are deductive." It would seem from the above somewhat singularly juxtaposed paragraphs that, according to Prof. Huxley, the business of the mathematical student is from a limited number of propositions (bottled up and labeled ready for future use) to deduce any required result by a process of the same general nature as a student of language employs in declining and conjugating his nouns and verbs—that to make out a mathematical proposition and to construe or parse a sentence are equivalent or identical mental operations. Such an opinion scarcely seems to need serious refutation. The passage is taken from an article in *Macmillan's Magazine* for June last, entitled "Scientific Education—Notes of an After-dinner Speech," and I cannot but think would have been couched in more guarded terms by my distinguished friend had his speech been made *before* dinner instead of *after*.

The notion that mathematical truth rests on the narrow basis of a limited number of elementary propositions from which all others are to be derived by a process of logical inference and verbal deduction, has been stated still more strongly and explicitly by the same eminent writer in an article of even date with the preceding in the *Fortnightly Review*, where we are told that "Mathematics is that study which knows nothing of observation, nothing of experiment, nothing of induction, nothing of causation." I think no statement could have been made more opposite to the undoubted facts of the case, that mathematical analysis is constantly invoking the aid of new principles, new ideas, and new methods, not capable of being defined by any form of words, but springing directly from the inherent powers and activity of the human mind, and from continually renewed introspection of that inner world of thought of which the phenomena are as varied and require as close attention to discern as those of the outer physical world (to which the inner one in each individual man may, I think, be conceived to stand in somewhat the same general relation of correspondence as a shadow to the object from which it is projected, or as the hollow palm of one hand to the closed fist which it grasps of the other), that it is unceasingly calling forth the faculties of observation and comparison, that one of its principal weapons is induction, that it has frequent recourse to experimental trial and verification, and that it affords a boundless scope for the exercise of the highest efforts of imagination and invention.

Lagrange, than whom no greater authority could be quoted, has expressed emphatically his belief in the importance to the mathematician of the faculty of observation; Gauss has called mathematics a science of the eye, and in conformity with this view always paid the most punctilious attention to preserve his text free from typographical errors; the ever to be lamented Riemann has written a thesis to show that the basis of our conception of space is purely empirical, and our knowledge of its laws the result of observation, that other kinds of space might be conceived to exist subject to laws different from those which govern the actual space in which we are immersed, and that there is no evidence of these laws extending to the ultimate infinitesimal elements of which space is composed. Like his master Gauss, Riemann refuses to accept Kant's doctrine of space and time being forms of intuition, and regards them as possessed of physical and objective reality. I may mention that Baron Sartorius von Waltershausen (a member of this Association) in his biography of Gauss ("Gauss Zu Gedächtniss"), published shortly after his death, relates that this great man used to say that he had laid aside several questions which he had treated analytically, and hoped to apply to them geometrical methods in a future state of existence, when his conceptions of space should have become amplified and extended; for as we can conceive beings (like infinitely attenuated bookworms in an infinitely thin sheet of paper) [7] which possess only the notion of space of two dimensions, so we may imagine beings capable of realizing space of four or a greater number of dimensions. [8] Our Cayley, the central luminary, the Darwin of the English school of mathematicians, started and elaborated at an early age, and with happy consequences, the same bold hypothesis.

Most, if not all, of the great ideas of modern mathematics have had their origin in observation. Take, for instance, the arithmetical theory of forms, of which the foundation was laid in the diophantine theorems of Fermat, left, without proof by their author, which resisted all the efforts of the myriad-minded Euler to reduce to demonstration, and only yielded up their cause of being when turned over in the blowpipe flame of Gauss's transcendent genius; or the doctrine of double periodicity, which resulted from the observation by Jacobi of a purely analytical fact of transformation; or Legendre's law of reciprocity; or Sturm's theorem about the roots of equations, which, as he informed me with his own lips, stared him in the face in the midst of some mechanical investigations connected with the motion of compound pendulums; or Huyghens' method of continued fractions, characterized by Lagrange as one of the principal discoveries of "that great mathematician, and to which he appears to have been led by the construction of his Planetary Automaton"; or the New Algebra, speaking of which one of my predecessors (Mr Spottiswoode) has said, not without just reason and authority, from this Chair, "that it reaches out and indissolubly connects itself each year with fresh branches of mathematics, that the theory of equations has almost become new through it, algebraic geometry transfigured in its light, that the calculus of variations, molecular physics, and mechanics" (he might, if speaking at the present moment, go on to add the theory of elasticity and the highest developments of the integral calculus) "have all felt its influence."

Now this gigantic outcome of modern analytical thought, itself, too, only the precursor and progenitor of a future, still more heaven-reaching theory, which will comprise a complete study of the interoperation, the actions and reactions, of algebraic forms (Analytical Morphology in its absolute sense), how did this originate? In the accidental observation by Eisenstein, some score or more years ago, of a single invariant (the Quadrinvariant of a Binary Quartic which he met within the course of certain researches just as accidentally and unexpectedly as M. Du Chaillu might meet a Gorilla in the country of the Fantees, or any one of us in London a White Polar Bear escaped from the Zoological Gardens. Fortunately, he pounced down upon his prey and preserved it for the contemplation and study of future mathematicians. It occupies only part of a page in his collected posthumous works. This single result of observation (as well entitled to be so called as the discovery of Globigerinrae in chalk or of the Confoco-ellipsoidal structure of the shells of the Foraminifera), which remained unproductive in the hands of its distinguished author, has served to set in motion a train of thought and to propagate an impulse which have led to a complete revolution in the whole aspect of modern analysis, and whose consequences will continue to be felt until Mathematics are forgotten and British Associations meet no more.

I might go on, were it necessary, piling instance upon instance to prove the paramount importance of the faculty of observation to the process of mathematical discovery. [9] Were it not unbecoming to dilate on one's personal experience, I could tell a story of almost romantic interest about my own latest researches in a field where Geometry, Algebra, and the Theory of Numbers melt in a surprising manner into one another, like sunset tints or the colors of the dying dolphin, "the last still loveliest" (a sketch of which has just appeared in the *Proceedings of the London Mathematical Society*), which would very strikingly illustrate how much observation, divination, induction, experimental trial, and verification, causation, too (if that means, as I suppose it must, mounting from phenomena to their reasons or causes of being), have to do with the work of the mathematician. In the face of these facts, which every analyst in this room or out of it can vouch for out of its own knowledge and personal experience, how can it be maintained, in the words of Professor Huxley, who, in this instance, is speaking of the sciences as they are in themselves and without any reference to scholastic discipline, that Mathematics "is that study which knows nothing of observation, nothing of induction, nothing of experiment, nothing of causation"?

I, of course, am not so absurd as to maintain that the habit of observation of external nature will be best or in any degree cultivated by the study of mathematics, at all events as that study is at present conducted; and no one can desire more earnestly than myself to see natural and experimental science introduced into our schools as a primary and indispensable branch of education: I think that that study and mathematical culture should go on hand in hand together, and that they would greatly influence each other for their mutual good. I should rejoice to see mathematics taught with that life and animation which the presence and example of her young and buoyant sister could not fail to impart, short roads preferred to long ones, Euclid honorably shelved or buried "deeper than did ever plummet sound" out of the school-boy's reach, morphology introduced into the elements of Algebra—projection, correlation, and motion accepted as aids to geometry—the mind of the student quickened and elevated and his faith awakened by early initiation into the ruling ideas of polarity, continuity, infinity, and familiarization with the doctrine of the imaginary and inconceivable.

It is this living interest in the subject which is so wanting in our traditional and mediaeval modes of teaching. In France, Germany, and Italy, everywhere where I have been on the Continent, mind acts direct on mind in a manner unknown to the frozen formality of our academic institutions schools of thought and centers of real intellectual cooperation exist the relation of master and pupil is acknowledged as a spiritual and a lifelong tie, connecting successive generations of great thinkers with each other in an unbroken chain, just in the same way as we read, in the catalogue of our French Exhibition, or of the Salon at Paris, of this man or that being the pupil of one great painter or sculptor and the master of another. When followed out in this spirit, there is no study in the world which brings into more harmonious action all the faculties of the mind than the one of which I stand here as the humble representative, there is none other which prepares so many agreeable surprises for its followers, more wonderful than the changes in the transformation-scene of a pantomime, or, like this, seems to raise them, by successive steps of initiation, to higher and higher states of conscious intellectual being.

This accounts, I believe, for the extraordinary longevity of all the greatest masters of the Analytical art, the Dei Majores of the mathematical Pantheon. Leibniz lived to the age of 70; Euler to 76; Lagrange to 77; Laplace to 78 ; Gauss to 78; Plato, the supposed inventor of the conic sections, who made mathematics his study and delight, who called them the handles or aids to philosophy, the medicine of the soul, and is said never to have let a day go by without inventing some new theorems, lived to 82; Newton, the crown and glory of his race, to 85 ; Archimedes, the nearest akin, probably, to Newton in genius, was 75, and might have lived on to be 100, for aught we can guess to the contrary, when he was slain by the impatient and ill-mannered sergeant, sent to bring him before the Roman general, in the full vigor of his faculties, and in the very act of working out a problem; Pythagoras, in whose school, I believe, the word mathematician (used, however, in a somewhat wider than its present sense) originated, the second founder of geometry, the inventor of the matchless theorem which goes by his name, the recognizer of the undoubtedly mis-called Copernican theory, the discoverer of the regular solids and the musical canon, who stands at the very apex of this pyramid of fame, (if we may credit the tradition) after spending 22 years studying in Egypt, and 12 in Babylon, opened school when 56 or 57 years old in Magna Græca, married a young wife when past 60, and died, carrying on his work with energy unspent to the last, at the age of 99. The mathematician lives long and lives young; the wings of his soul do not early drop off, nor do its pores become clogged with the earthy particles blown from the dusty highways of vulgar life.

Some people have been found to regard all mathematics, after the 47th proposition of Euclid, as a sort of morbid secretion, to be compared only with the pearl said to be generated in the diseased oyster, or, as I have heard it described, "une excroissance maladive de l'esprit humain." Others find its justification, its "raison d'être," in its being either the torch-bearer leading the way, or the handmaiden holding up the train of Physical Science; and a very clever writer in a recent magazine article, expresses his doubts whether it is, in itself, a more serious pursuit, or more worthy of interesting an intellectual human being, than the study of chess problems or Chinese puzzles. What is it to us, they say, if the three angles of a triangle are equal to two right angles, or if every even number is, or may be, the sum of two primes, or if every equation of an odd degree must have a real root. How dull, stale, flat, and unprofitable are such and such like announcements! Much more interesting to read an account of a marriage in high life, or the details of an international boat-race. But this is like judging of architecture from being shown some of the brick and mortar, or even a quarried stone of a public building, or of painting from the colors mixed on the palette, or of music by listening to the thin and screechy sounds produced by a bow passed haphazard over the strings of a violin. The world of ideas which it discloses or illuminates, the contemplation of divine beauty and order which it induces, the harmonious connection of its parts, the infinite hierarchy and absolute evidence of the truths with which it is concerned, these, and such like, are the surest grounds of the title of mathematics to human regard, and would remain unimpeached and unimpaired were the plan of the universe unrolled like a map at our feet, and the mind of man qualified to take in the whole scheme of creation at a glance.

In conformity with general usage, I have used the word mathematics in the plural; but I think it would be desirable that this form of word should be reserved for the applications of the science, and that we should use mathematic in the singular number to denote the science itself, in the same way as we speak of logic, rhetoric, or (own sister to algebra) music. Time was when all the parts of the subject were dissevered, when algebra, geometry, and arithmetic either lived apart or kept up cold relations of acquaintance confined to occasional calls upon one another; but that is now at an end; they are drawn together and are constantly becoming more and more intimately related and connected by a thousand fresh ties, and we may confidently look forward to a time when they shall form but one body with one soul. Geometry formerly was the chief borrower from arithmetic and algebra, but it has since repaid its obligations, with abundant usury; and if I were asked to name, in one word, the pole-star round which the mathematical firmament revolves, the central idea which pervades as a hidden spirit the whole corpus of mathematical doctrine, I should point to Continuity as contained in our notions of space, and say, it is this, it is this! Space is the *Grand Continuum* from which, as from an inexhaustible reservoir, all the fertilizing ideas of modern analysis are derived; and as Brindley, the engineer, once allowed before a parliamentary committee that, in his opinion, rivers were made to feed navigable canals, I feel almost tempted to say that one principal reason for the existence of space, or at least one principal function which it discharges, is that of feeding mathematical invention. Everybody knows what a wonderful influence geometry has exercised in the hands of Cauchy, Puiseux, Riemann, and his followers Clebsch, Gordan, and others, over the very form and presentment of the modern calculus, and how it has come to pass that the tracing of curves, which was once to be regarded as a puerile amusement, or at best useful only to the architect or decorator, is now entitled to take rank as a high philosophical exercise, inasmuch as every new curve or surface, or other circumscription of space is capable of being regarded as the embodiment of some specific organized system of continuity.[10]

The early study of Euclid made me a hater of Geometry, which I hope may plead my excuse if I have shocked the opinions of any in this room (and I know there are some who rank Euclid as second in sacredness to the Bible alone, and as one of the advanced outposts of the British Constitution) by the tone in which I have previously alluded to it as a school-book; and yet, in spite of this repugnance, which had become a second nature in me, whenever I went far enough into any mathematical question, I found I touched, at last, a geometrical bottom: so it was, I may instance, in the purely arithmetical theory of partitions; so, again, in one of my more recent studies, the purely algebraical question of the invariantive criteria of the nature of the roots of an equation of the fifth degree: the first inquiry landed me in a new theory of polyhedra; the latter found its perfect and only possible complete solution in the construction of a surface of the ninth order and the subdivision of its infinite content into three distinct natural regions.

Having thus expressed myself at much greater length than I originally intended on the subject, which, as standing first on the muster-roll of the Association, and as having been so recently and repeatedly arraigned before the bar of public opinion, is entitled to be heard in its defense (if anywhere) in this place—having endeavored to show what it is not, what it is, and what it is probably destined to become, I feel that I must enough and more than enough have trespassed on your forbearance, and shall proceed with the regular business of the Meeting.

Before calling upon the authors of the papers contained in the varied bill of intellectual fare which I see before me, I hope to be pardoned if I direct attention to the importance of practicing brevity and condensation in the delivery of communications to the Section, not merely as a saving of valuable time, but in order that what is said may be more easily followed and listened to with greater pleasure and advantage. I believe that immense good may be done by the oral interchange and discussion of ideas which takes place in the Sections; but for this to be possible, details and long descriptions should be reserved for printing and reading, and only the general outlines and broad statements of facts, methods, observations, or inventions brought before us here, such as can be easily followed by persons having a fair average acquaintance with the several subjects treated upon. I understand the rule to be that, with the exception of the author of any paper who may answer questions and reply at the end of the discussion, no member is to address the Section more than once on the same subject, or occupy more than a quarter of an hour in speaking.

In order to get through the business set down in each day's paper, it may sometimes be necessary for me to bring a discussion to an earlier close than might otherwise be desirable, and for that purpose to request the authors of papers, and those who speak upon them, to be brief in their addresses. I have known most able investigators at these Meetings, and especially in this Section, gradually part company with their audience, and at last become so involved in digressions as to lose entirely the thread of their discourse, and seem to forget, like men waking out of sleep, where they were or what they were talking about. In such cases I shall venture to give a gentle pull to the string of the kite before it soars right away out of sight into the region of the clouds. I now call upon Dr

Magnus to read his paper and recount to the Section his wondrous story on the Emission, Absorption, and Reflection of Obscure Heat. [11]

POSTSCRIPT—The remarks on the use of experimental methods in mathematical investigation led to Dr Jacobi, the eminent physicist of St Petersburg, who was present at the delivery of the address, favoring me with the annexed anecdote relative to his illustrious brother C.G. J. Jacobi.

"En causant un jour avec mon frère défunt sur la necessité de contrôler par des expériences réiterées toute observation, même si elle confirme l'hypothèse, il me raconta avoir découvert un jour une loi très remarquable de la théorie des nombres, dont il ne douta guère qu'elle fut générale. Cependant par un excès de précaution ou plutôt pour faire le superflu, il voulut substituer un chiffre quelconque réel aux termes généraux, chiffre qu'il choisit au hasard ou, peut-être, par une espèce de divination, car en effet ce chiffre mit sa formule en défaut; tout autre chiffre qu'il essaya en confirma la généralité. Plus tard il réussit à prouver que le chiffre choisi par lui par hasard, appartenait à un système de chiffres qui faisait la seule exception à la règle.

"Ce fait curieux m'est resté dans la mémoire, mais comme il s'est passé il y a plus d'une trentaine d'années, je ne rappelle plus des détails.

"M. H. JACOBI."

"EXETER, 24. *Août,* 1869."

[1] The Address was also reprinted by the author in a volume issued by Longmans, Green and Co., London, 1870, of which the earlier portion deals with the Laws of Verse.

[2] So it is said of Jacobi, that he attracted the particular attention and friendship of Böckh, the director of the philological seminary at Berlin, by the zeal and talent he displayed for philology, and only at the end of two years' study at the University, and after a severe mental struggle, was able to make his final choice in favor of mathematics. The relation between these two sciences is not perhaps so remote as may at first sight appear, and indeed it has often struck me that metamorphosis runs like a golden thread through the most diverse branches of modern intellectual culture, and forms a natural link of connection between subjects in their aims so unlike as Grammar, Ethnology, Rational Mythology, Chemistry, Botany, Comparative Anatomy, Physiology, Physics, Algebra, Music, all of which, under the modern point of view, may be regarded as having morphology for their common center. Even singing, I have been told, the advanced German theorists regard as being strictly a development of recitative, and infer therefrom that no essentially new melodic themes can be invent-

ed until a social cataclysm, or the civilization of some at present barbaric races, shall have created new necessities of expression and called into activity new forms of impassioned declamation.

[3] My first printed paper was on Fresnel's Optical Theory, published in the *Philosophical Magazine*; my latest contribution to the *Philosophical Transactions* is a memoir on the "Rotation of a Free Rigid Body." There is an old adage, "purus mathematicus, purus asinus." On the other hand, I once heard the great Richard Owen say, when we were opposite neighbors in Lincoln's-Inn Fields (doves nestling among hawks), that he would like to see *Homo Mathematicus* constituted into a distinct subclass, thereby suggesting to my mind sensation, perception, reflection, abstraction, as the successive stages or phases of protoplasm on its way to being made perfect Mathematicised Man. Would it sound too presumptuous to speak of perception as a quintessence of sensation, language (that is, communicable thought) of perception, mathematic of language? We should then have four terms differentiating from inorganic matter and from each other the Vegetable, Animal, Rational, and Supersensual modes of existence.

[4] Mr Spottiswoode favored the Section, in his opening address, with a combined history of the progress of Mathematics and Physics; Dr Tyndall's address was virtually on the limits of Physical Philosophy; the one here in print is an attempted faint adumbration of the nature of Mathematical Science in the abstract. What is wanting (like a fourth sphere resting on three others in contact) to build up the Ideal Pyramid is a discourse on the Relation of the two branches (Mathematics and Physics) to their action and reaction upon one another, a magnificent theme, with which it is to be hoped some future President of Section A will crown the edifice and make the Tetralogy symbolizable by $A + A', A, A', A \cdot A'$ complete.

[5] Although no great lecture-goer, I have heard three lectures in my life which have left a lasting impression as masterpieces on my memory—Clifford on Mind, Huxley on Chalk, Dumas on Faraday.

[6] In his *éloge* of Daubenton, Cuvier remarks, "Les savants jugent toujours comme le vulgaire les ouvrages qui ne sont pas de leur genre."

[7] I have read or been told that eye of observer has never lighted on these depredators, living or dead. Nature has gifted me with eyes of exceptional microscopic power, and I can speak with some assurance of having repeatedly seen the creature wriggling on the learned page. On approaching it with breath or fingernail it stiffens out into the semblance of a streak of dirt, and so eludes detection.

[8] It is well known to those who have gone into these views that the laws of motion accepted as a fact suffice to prove in a general way that the space we live in is a flat or level space (a "homaloid"), our existence therein being assimilable to the life of the bookworm in an *unrumpled page:* but what if the page should be undergoing a process of gradual bending into a curved form? Mr. W. K. Clifford has indulged in some remarkable speculations as to the possibility of our being able to infer, from certain unexplained phenomena of light and

magnetism, the fact of our level space of three dimensions being in the act of undergoing in space of four dimensions (space as inconceivable to us as our space to the supposititious bookworm) a distortion analogous to the rumpling of the page to which that creature's powers of direct perception have been postulated to be limited.

[9] Newton's Rule was to all appearance, and according to the more received opinion, obtained inductively by its author. My own reduction of Euler's problem of the Virgins (or rather one slightly more general than this) to the form of a question (or, to speak more exactly, a set of questions) in simple partitions was, strange to say, first obtained by myself inductively, the result communicated to Prof. Cayley, and proved subsequently by each of us independently, and by perfectly distinct methods.

[10] M. Camille Jordan's application of Dr Salmon's Eikosi-heptagram to Abelian functions is one of the most recent instances of this reverse action of geometry on analysis. Mr Crofton's admirable apparatus of a reticulation with infinitely fine meshes rotated successively through indefinitely small angles, which he applies to obtaining whole families of definite integrals, is another equally striking example of the same phenomenon.

[11] Curiously enough, and as if symptomatic of the genial warmth of the proceedings in which seven sages from distant lands (Jacobi, Magnus, Newton, Janssen, Morren, Lyman, Neumayer) took frequent part, the opening and concluding papers (each of surpassing interest, and a letting-out of mighty waters) were on Obscure Heat, by Prof. Magnus, and on Stellar Heat, by Mr. Huggins.

From the intrinsic evidence of his creation, the Great Architect of the Universe now begins to appear as a pure mathematician. — James Jeans (1877–1946)

The Mathematician

John von Neumann

János von Neumann (1903–1957), or "Johnny" as he was later to be called in America, was a child prodigy, who in his relatively short life made significant and enduring contributions to a wide range of topics—mathematics, physics, computer engineering, and mathematical economics.

He was born in Budapest, the son of Miksa (Max) Neumann and Margit (Margaret) Kann. The parents were members of the affluent Jewish community of Budapest—affluent both financially and intellectually. Margaret's father was a very successful businessman who, among other things, copied in Hungary, the successful practices of Sears Roebuck. We also explain parenthetically, that in 1913 Max was given the privilege of adding the honorific "von" to his name but he did not avail himself of that option. His son John however did and used the German form "von Neumann." In addition to János, the parents had two other children—Nicholas and Michael.

Von Neumann had a phenomenal memory and as a child could read classical Greek and perform prodigious mathematical calculations in his head. The environment in which he was reared offered all possible educational advantages—governesses who taught him German and French, for example. Dinner in the Neumann household was a constant source of intellectual stimulation.

János attended one of the more demanding secondary schools but, after graduation, his father was skeptical of a career in mathematics as one having limited financial opportunity. With the help of Theodore von Kármán, he urged János to study chemistry and eventually von Neumann got a diploma in chemical engineering from Zürich in 1925. The lure of mathematics however was too great and, accordingly, he completed a doctoral degree in mathematics in Budapest in 1928. While in Zürich he had stimulating contacts with Hermann Weyl and George Pólya.

In 1930, von Neumann married Mariette Korvesi and at the same time the family converted to Catholicism. The Neumanns, however, were not active either in Judaism or in Catholicism. On the other hand, von Neumann, giving in to Pascal's wager on his death bed, received extreme unction.

John and Marietta had one daughter Marina but were divorced in 1936. In 1938 he married Klari Dan.

His mathematical achievements very quickly attracted wide attention; he lectured in Berlin between 1926 and1929 and accepted a position as visiting lecturer at Princeton in 1930. When the Institute for Advanced Study was established in Princeton in 1933, he was one of the six professors appointed. He was a very lively host and spirited conversationalist and the von Neumann home was a place of great hospitality. Although he continued his European contacts, the coming to power of the Nazis induced him to sever all ties with National Socialist Germany.

During and after World War II he served as a consultant to the U.S. armed services. In one of these roles his valuable contributions included the "implosion" method for bringing nuclear fuel to critical mass, thus hastening the development of the atomic bomb. After the war he continued as a consultant to the government and espoused what were, at that time, called "hawkish" policies. His cold-war political views were not popular with many colleagues.

In his research activities, he had become interested in, among other things, hydrodynamical turbulence and in the analysis of the underlying non-linear partial differential equations. Numerical analysis seemed to be the only path to insights into this difficult field and the need to perform elaborate calculations impelled him to study new techniques for performing these calculations and, in particular, to investigate the burgeoning field of electronic computers. His ideas contributed significantly to the development of techniques and methodologies currently in use. He presided over the construction of an electronic computer in Princeton.

His wide-ranging intellect led him to undertake research in economics and with Oskar Morgenstern, Neumann wrote a seminal work on mathematical economics.

Tragedy struck when in 1956 he contracted incurable cancer and he died in torment the following year.

During his lifetime he was showered with honors from universities and academies all over the world. In addition he received Presidential awards as well as the Fermi and Einstein awards.

He left a rich legacy which, like the amaranth, will not fade for many years to come.

Editor's Preface

In this essay, von Neumann discusses the nature of an intellectual discipline in general, and mathematics in particular. He notes that this would be difficult for any field of endeavor but is especially so in mathematics.

Over the centuries, mathematicians have developed symbolism as a highly economical sort of shorthand with which to communicate ideas. This symbolism has evolved to such an extent that it is difficult, if not well nigh impossible, to convey to the nonspecialist any but the vaguest idea of the underlying content. This telegraphic quality is, however, one of the strengths of the subject. Indeed it is the mechanism by which deep ideas can be made relatively accessible. This shorthand has enabled us to penetrate deeply into the mysteries of mathematics. The absence of such economy of thought can be an impediment to progress. For example, a noted mathematician, B.L. Van der Waerden, has surmised that the main reason for the decline of Euclidean geometry was the increasing complexity and intricacy of the figures with which students and geometers had to deal. Other disciplines—scientific or otherwise, share this linguistic and conceptual difficulty; it is likely the case, however, that in the sciences, the difficulty is roughly proportional to the extent to which the science has been mathematized.

One of the questions addressed by von Neumann is the question: Is mathematics empirical? In addressing this question, we need to elaborate on the meaning of the term "empirical." It can be interpreted in two senses:

(1) Is the invention of the mathematical concept directed at the solution of a "real" world problem such as the area of an irregularly shaped farm abutting the river Nile (as in Egyptian mathematics)? To give another example, is it aimed at the solution to the question of determining the longitude of a point on the surface of the earth?

(2) Or is it a pursuit which finds its inspiration *indirectly* in the "real" world. Such an example might be the question of determining all simple groups. Other examples, however, can readily be cited.

In the second case, mathematics becomes more akin to the creations of artists—only the tools differ. The artist finds inspiration from "real world" objects but creates an imaginary composition such as, for example, Raphael's "School of Athens." In the realm of music, the composer can integrate real world sounds into a musical composition.

Von Neumann leans toward the first view but expresses concern about the logical basis of empiricism. He also expresses anxiety over mathematical pursuits that are not firmly rooted in empirical sources since, if a mathematical discipline strays too far from reality, he warns, it runs the risk of degenerating into a sterile inquiry leading to contrived and sterile forms of structures. The history of mathematical ideas suggests otherwise. To counter this uneasiness, one can point to many examples of abstract mathematical constructs which, long after their creation, find expression in reality. Von Neumann himself is one such creator! Could it be that the brain stores a real image as an abstract neural network? We leave such speculations to neurologists.

A discussion of the nature of intellectual work is a difficult task in any field, even in fields which are not so far removed from the central area of our common human intellectual effort as mathematics still is. A discussion of the nature of any intellectual effort is difficult *per se*—at any rate, more difficult than the mere exercise of that particular intellectual effort. It is harder to understand the mechanism of an airplane, and the theories of the forces which lift and which propel it, than merely to ride in it, to be elevated and transported by it or even to steer it. It is exceptional that one should be able to acquire the understanding of a process without having previously acquired a deep familiarity with running it, with using it, before one has assimilated it in an instinctive and empirical way.

Thus any discussion of the nature of intellectual effort in any field is difficult, unless it presupposes an easy, routine familiarity with that field. In mathematics this limitation becomes very severe, if the discussion is to be kept on a nonmathematical plane. The discussion will then necessarily show some very bad features; points which are made can never be properly documented, and a certain over-all superficiality of the discussion becomes unavoidable.

I am very much aware of these shortcomings in what I am going to say, and I apologize in advance. Besides, the views which I am going to express are probably not wholly shared by many other mathematicians —you will get one man's not-too-well systematized impressions and interpretations—and I can give you only very little help in deciding how much they are to the point.

In spite of all these hedges, however, I must admit that it is an interesting and challenging task to make the attempt and to talk to you about the nature of intellectual effort in mathematics. I only hope that I will not fail too badly.

The most vitally characteristic fact about mathematics is, in my opinion, its quite peculiar relationship to the natural sciences, or more generally, to any science which interprets experience on a higher than purely descriptive level.

Most people, mathematicians and others, will agree that mathematics is not an empirical science, or at least that it is practiced in a manner which differs in several decisive respects from the techniques of the empirical sciences. And, yet, its development is very closely linked with the natural sciences. One of its main branches, geometry, actually started as a natural, empirical science. Some of the best inspirations of modern mathematics (I believe, the best ones) clearly originated in the natural sciences. The methods of mathematics pervade and dominate the "theoretical" divisions of the natural sciences. In modern empirical sciences it has become more and more a major criterion of success whether they have become accessible to the mathematical method or to the near-mathematical methods of physics. Indeed, throughout the natural sciences an unbroken chain of successive pseudomorphoses, all of them pressing toward mathematics, and almost identified with the idea of scientific progress, has become more and more evident. Biology becomes increasingly pervaded by chemistry and physics, chemistry by experimental and theoretical physics, and physics by very mathematical forms of theoretical physics.

There is a quite peculiar duplicity in the nature of mathematics. One has to realize this duplicity, to accept it, and to assimilate it into one's thinking on the subject. This double face is the face of mathematics, and I do not believe that any simplified, unitarian view of the thing is possible without sacrificing the essence.

I will therefore not attempt to present you with a unitarian version. I will attempt to describe, as best I can, the multiple phenomenon which is mathematics.

It is undeniable that some of the best inspirations in mathematics—in those parts of it which are as pure mathematics as one can imagine have come from the natural sciences. We will mention the two most monumental facts.

The first example is, as it should be, geometry. Geometry was the major part of ancient mathematics. It is, with several of its ramifications, still one of the main divisions of modern mathematics. There can be no doubt that its origin in antiquity was empirical and that it began as a discipline not unlike theoretical physics today. Apart from all other evidence, the very name "geometry" indicates this. Euclid's postulational treatment represents a great step away from empiricism, but it is not at all simple to defend the position that this was the decisive and final step, producing an absolute separation. That Euclid's axiomatization does at some minor points not meet the modern requirements of absolute axiomatic rigor is of lesser importance in this respect. What is more essential, is this: other disciplines, which are undoubtedly empirical, like mechanics and thermodynamics, are usually presented in a more or less postulational treatment, which in the presentation of some authors is hardly distinguishable from Euclid's procedure. The classic of theoretical physics in our time, Newton's Principia, was, in literary form as well as in the essence of some of its most critical parts, very much like Euclid. Of course in all these instances there is behind the postulational presentation the physical insight backing the postulates and the experimental verification supporting the theorems. But one might well argue that a similar interpretation of Euclid is possible, especially from the viewpoint of antiquity, before geometry had acquired its present bimillenial stability and authority—an authority which the modern edifice of theoretical physics is clearly lacking.

Furthermore, while the de-empirization of geometry has gradually progressed since Euclid, it never became quite complete, not even in modern times. The discussion of non-Euclidean geometry offers a good illustration of this. It also offers an illustration of the ambivalence of mathematical thought. Since most of the discussion took place on a highly abstract plane, it dealt with the purely logical problem whether the "fifth postulate" of Euclid was a consequence of the others or not; and the formal conflict was terminated by F. Klein's purely mathematical example, which showed how a piece of a Euclidean plane could be made non-Euclidean by formally redefining certain basic concepts. And yet the empirical stimulus was there from start to finish. The prime reason, why, of all Euclid's postulates, the fifth was questioned, was clearly the unempirical character of the concept of the entire infinite plane which intervenes there, and there only. The idea that in at least one significant sense—and in spite of all mathematico-logical analyses—the decision for or against Euclid may have to be empirical, was certainly present in the mind of the greatest mathematician, Gauss. And after Bolyai, Lobatschefski, Riemann, and Klein had obtained *more abstracto,* what we today consider the formal resolution of the original controversy, empirics—or rather physics—nevertheless, had the final say. The discovery of general relativity forced a revision of our views on the relationship of geometry in an entirely new setting and with a quite new distribution of the purely mathematical emphases, too. Finally, one more touch to complete the picture of contrast. This last development took place in the same generation which saw the complete de-empirization and abstraction of Euclid's axiomatic method in the hands of the modern axiomatic-logical mathematicians. And these two seemingly conflicting attitudes are perfectly compatible in one mathematical mind; thus Hilbert made important contributions to both axiomatic geometry and to general relativity.

The second example is calculus—or rather all of analysis, which sprang from it. The calculus was the first achievement of modern mathematics, and it is difficult to overestimate its importance. I think it defines more unequivocally than anything else the inception of modern mathematics, and the system of mathematical analysis, which is its logical development, still constitutes the greatest technical advance in exact thinking.

The origins of calculus are clearly empirical. Kepler's first attempts at integration were formulated as "dolichometry"—measurement of kegs, that is, volumetry for bodies with curved surfaces. This is geometry, but post-Euclidean, and, at the epoch in question, nonaxiomatic, empirical geometry. Of this, Kepler was fully aware. The main effort and the main discoveries, those of Newton and Leibniz, were of an explicitly physical origin. Newton invented the calculus "of fluxions" essentially for the purposes of mechanics—in fact, the two disciplines, calculus and mechanics, were developed by him more or less together. The first formulations of the calculus were not even mathematically rigorous. An inexact, semiphysical formulation was the only one available for over a hundred and fifty years after Newton! And yet, some of the most important advances of analysis took place during this period, against this inexact, mathematically inadequate background! Some of the leading mathematical spirits of the period were clearly not rigorous, like Euler; but others, in the main, were, like Gauss or Jacobi. The development was as confused and ambiguous as can be, and its relation to empiricism was certainly not according to our present (or Euclid's) ideas of abstraction and rigor. Yet no mathematician would want to exclude it from the fold—that period produced mathematics as first class as ever existed! And even after the reign of rigor was essentially re-established with Cauchy, a very peculiar relapse into semiphysical methods took place with Riemann. Riemann's scientific personality itself is a most illuminating example of the double nature of mathematics, as is the controversy of Riemann and Weierstrass, but it would take me too far into technical matters if I went into specific details. Since Weierstrass, analysis seems to have become completely abstract, rigorous, and unempirical. But even this is not unqualifiedly true. The controversy about the "foundations" of mathematics and logics, which took place during the last two generations, dispelled many illusions on this score.

This brings me to the third example which is relevant for the diagnosis. This example, however, deals with the relationship of mathematics with philosophy or epistemology rather than with the natural sciences. It illustrates, in a very striking fashion, that the very concept of "absolute" mathematical rigor is not immutable. The variability of the concept of rigor shows that something else besides mathematical abstraction must enter into the makeup of mathematics. In analyzing the controversy about the "foundations," I have not been able to convince myself that the verdict must be in favor of the empirical nature of this extra component. The case in favor of such an interpretation is quite strong, at least in some phases of the discussion. But I do not consider it absolutely cogent. Two things, however, are clear. First, that something nonmathematical, somehow connected with the empirical sciences or with philosophy or both, does enter essentially and its nonempirical character could only be maintained if one assumed that philosophy (or more specifically epistemology) can exist independently of experience. (And this assumption is only necessary but not in itself sufficient). Second, that the empirical origin of mathematics is strongly supported by instances like our two earlier examples (geometry and calculus), irrespective of what the best interpretation of the controversy about the "foundations" may be.

In analyzing the variability of the concept of mathematical rigor, I wish to lay the main stress on the "foundations" controversy, as mentioned above. I would, however, like to consider first briefly a secondary aspect of the matter. This aspect also strengthens my argument, but I do consider it as secondary, because it is probably less conclusive than the analysis of the "foundations" controversy. I am referring to the changes of mathematical "style." It is well known that the style in which mathematical proofs are written has undergone considerable fluctuations. It is better to talk of fluctuations than of a trend because in some respects the difference between the present and certain authors of the eighteenth or of the nineteenth century is greater than between the present and Euclid. On the other hand, in other respects there has been remarkable constancy. In fields in which differences are present, they are mainly differences in presentation, which can be eliminated without bringing in any new ideas. However, in many cases these differences are so wide that one begins to doubt whether authors who "present their cases" in such divergent ways can have been separated by differences in style, taste, and education only—whether they can really have had the same ideas as to what constitutes mathematical rigor. Finally, in the extreme cases (e.g., in much of the work of the late-eighteenth-century analysis, referred to above), the differences are essential and can be remedied, if at all, only with the help of new and profound theories, which it took up to a hundred years to develop. Some of the mathematicians who worked in such, to us, unrigorous ways (or some of their contemporaries, who criticized them) were well aware of their lack of rigor. Or to be more objective: their own desires as to what mathematical procedure should be were more in conformity with our present views than their actions. But others—the greatest virtuoso of the period, for example, Euler—seem to have acted in perfect good faith and to have been quite satisfied with their own standards.

However, I do not want to press this matter further. I will turn instead to a perfectly clear-cut case, the controversy about the "foundations of mathematics." In the late nineteenth and the early twentieth centuries a new branch of abstract mathematics, G. Cantor's theory of sets, led into difficulties. That is, certain reasonings led to contradiction; and, while these reasonings were not in the central and "useful" part of set theory, and always easy to spot by certain formal criteria, it was nevertheless not clear why they should be deemed less set-theoretical than the "successful" parts of the theory. Aside from the *ex post* insight that they actually led into disaster, it was not clear what *a priori* motivation, what consistent philosophy of the situation, would permit one to segregate them from those parts of set theory which one wanted to save. A closer study of the *merita* of the case, undertaken mainly by Russell and Weyl, and concluded by Brouwer, showed that the way in which not only set theory but also most of modern mathematics used the concepts of "general validity" and of "existence" was philosophically objectionable. A system of mathematics which was free of these undesirable traits, "intuitionism," was developed by Brouwer. In this system the difficulties and contradiction of set theory did not arise. However, a good fifty per cent of modern mathematics, in its most vital and up to then unquestioned—parts, especially in analysis, were also affected by this "purge"; they either became invalid or had to be justified by very complicated subsidiary considerations. And in this latter process one usually lost appreciably in generality of validity and elegance of deduction. Nevertheless, Brouwer and Weyl considered it necessary that the concept of mathematical rigor be revised according to these ideas.

It is difficult to overestimate the significance of these events. In the third decade of the twentieth century two mathematicians—both of them of the first magnitude, and as deeply and fully conscious of what mathematics is, or is for, or is about, as anybody could be—actually proposed that the concept of mathematical rigor, of what constitutes an exact proof, should be changed! The developments which followed are equally worth noting.

1. Only very few mathematicians were willing to accept the new, exigent standards for their own daily use. Very many, however, admitted that Weyl and Brouwer were prima facie right, but they themselves continued to trespass, that is, to do their own mathematics in the old, "easy" fashion—probably in the hope that somebody else, at some other time, might find the answer to the intuitionistic critique and thereby justify them *a posteriori*.
2. Hilbert came forward with the following ingenious idea to justify "classical" (i.e., pre-intuitionistic) mathematics: Even in the intuitionistic system it is possible to give a rigorous account of how classical mathematics operate, that is, one can describe how the classical system works, although one cannot justify its workings. It might therefore be possible to demonstrate intuitionistically that classical procedures can never lead into contradictions—into conflicts with each other. It was clear that such a proof would be very difficult, but

there were certain indications how it might be attempted. Had this scheme worked, it would have provided a most remarkable justification of classical mathematics on the basis of the opposing intuitionistic system itself! At least, this interpretation would have been legitimate in a system of the philosophy of mathematics which most mathematicians were willing to accept.

3. After about a decade of attempts to carry out this program, Gödel produced a most remarkable result. This result cannot be stated absolutely precisely without several clauses and caveats which are too technical to be formulated here. Its essential import, however, was this: If a system of mathematics does not lead into contradiction, then this fact cannot be demonstrated with the procedures of that system. Gödel's proof satisfied the strictest criterion of mathematical rigor—the intuitionistic one. Its influence on Hilbert's program is somewhat controversial, for reasons which again are too technical for this occasion. My personal opinion, which is shared by many others, is, that Gödel has shown that Hilbert's program is essentially hopeless.
4. The main hope of a justification of classical mathematics—in the sense of Hilbert or of Brouwer and Weyl—being gone, most mathematicians decided to use that system anyway. After all, classical mathematics was producing results which were both elegant and useful, and, even though one could never again be absolutely certain of its reliability, it stood on at least as sound a foundation as, for example, the existence of the electron. Hence, if one was willing to accept the sciences, one might as well accept the classical system of mathematics. Such views turned out to be acceptable even to some of the original protagonists of the intuitionistic system. At present the controversy about the "foundations" is certainly not closed, but it seems most unlikely that the classical system should be abandoned by any but a small minority.

I have told the story of this controversy in such detail, because I think that it constitutes the best caution against taking the immovable rigor of mathematics too much for granted. This happened in our own lifetime, and I know myself how humiliatingly easily my own views regarding the absolute mathematical truth changed during this episode, and how they changed three times in succession!

I hope that the above three examples illustrate one-half of my thesis sufficiently well—that much of the best mathematical inspiration comes from experience and that it is hardly possible to believe in the existence of an absolute, immutable concept of mathematical rigor, dissociated from all human experience. I am trying to take a very low-brow attitude on this matter. Whatever philosophical or epistemological preferences anyone may have in this respect, the mathematical fraternities' actual experiences with its subject give little support to the assumption of the existence of an *a priori* concept of mathematical rigor. However, my thesis also has a second half, and I am going to turn to this part now.

It is very hard for any mathematician to believe that mathematics is a purely empirical science or that all mathematical ideas originate in empirical subjects. Let me consider the second half of the statement first. There are various important parts of modern mathematics in which the empirical origin is untraceable, or, if traceable, so remote that it is clear that the subject has undergone a complete metamorphosis since it was cut off from its empirical roots. The symbolism of algebra was invented for domestic, mathematical use, but it may be reasonably asserted that it had strong empirical ties. However, modern, "abstract" algebra has more and more developed into directions which have even fewer empirical connections. The same may be said about topology. And in all these fields the mathematician's subjective criterion of success, of the worth-whileness of his effort, is very much self-contained and aesthetical and free (or nearly free) of empirical connections. (I will say more about this further on.) In set theory this is still clearer. The "power" and the "ordering" of an infinite set may be the generalizations of finite numerical concepts, but in their infinite form (especially "power") they have hardly any relation to this world. If I did not wish to avoid technicalities, I could document this with numerous set theoretical examples—the problem of the "axiom of choice," the "comparability" of infinite "powers," the "continuum problem," etc. The same remarks apply to much of real function theory and real point-set theory. Two strange examples are given by differential geometry and by group theory: they were certainly conceived as abstract, nonapplied disciplines and almost always cultivated in this spirit. After a decade in one case, and a century in the other, they turned out to be very useful in physics. And they are still mostly pursued in the indicated. abstract, nonapplied spirit.

The examples for all these conditions and their various combinations could be multiplied, but I prefer to turn instead to the first point I indicated above: Is mathematics an empirical science? Or, more precisely:

Is mathematics actually practiced in the way in which an empirical science is practiced? Or, more generally: What is the mathematician's normal relationship to his subject? What are his criteria of success, of desirability? What influences, what considerations, control and direct his effort?

Let us see, then, in what respects the way in which the mathematician normally works differs from the mode of work in the natural sciences. The difference between these, on one hand, and mathematics, on the other, goes on, clearly increasing as one passes from the theoretical disciplines to the experimental ones and then from the experimental disciplines to the descriptive ones. Let us therefore compare mathematics with the category which lies closest to it—the theoretical disciplines. And let us pick there the one which lies closest to mathematics. I hope that you will not judge me too harshly if I fail to control the mathematical *hubris* and add: because it is most highly developed among all theoretical sciences—that is, theoretical physics. Mathematics and theoretical physics have actually a good deal in common. As I have pointed out before, Euclid's system of geometry was the prototype of the axiomatic presentation of classical mechanics, and similar treatments dominate phenomenological thermodynamics as well as certain phases of Maxwell's system of electrodynamics and also of special relativity. Furthermore, the attitude that theoretical physics does not explain phenomena, but only classifies and correlates, is today accepted by most theoretical physicists. This means that the criterion of success for such a theory is simply whether it can, by a simple and elegant classifying and correlating scheme, cover very many phenomena, which without this scheme would seem complicated and heterogeneous, and whether the scheme even covers phenomena which were not considered or even not known at the time when the scheme was evolved. (These two latter statements express, of course, the unifying and the predicting power of a theory.) Now this criterion, as set forth here, is clearly to a great extent of an aesthetical nature; for this reason it is very closely akin to the mathematical criteria of success, which, as you shall see, are almost entirely aesthetical. Thus we are now comparing mathematics with the empirical science that lies closest to it and with which it has, as I hope I have shown, much in common—with theoretical physics. The differences in the actual *modus procedendi* are nevertheless great and basic. The aims of theoretical physics are in the main given from the "outside," in most cases by the needs of experimental physics. They almost always originate in the need of resolving a difficulty; the predictive and unifying achievements usually come afterward. It we may be permitted a simile, the advances (predictions and unifications) come during the pursuit, which is necessarily preceded by a battle against some pre-existing difficulty (usually an apparent contradiction within the existing system). Part of the theoretical physicists's work is a search for such obstructions, which promise a possibility for a "break-through." As I mentioned, these difficulties originate usually in experimentation, but sometimes they are contradictions between various parts of the accepted body of theory itself. Examples are, of course, numerous.

Michelson's experiment leading to special relativity, the difficulties of certain ionization potentials and of certain spectroscopic structures leading to quantum mechanics exemplify the first case; the conflict between special relativity and Newtonian gravitational theory leading to general relativity exemplifies the second, rarer case. At any rate, the problems of theoretical physics are objectively given; and, while the criteria which govern the exploration of a success are, as I indicated earlier, mainly aesthetical, yet the portion of the problem, and that which I called above the original "break through," are hard, objective facts. Accordingly, the subject of theoretical physics was at almost all times enormously concentrated; at almost all times most of the effort of all theoretical physicists was concentrated on no more than one or two very sharply circumscribed fields, quantum theory in the 1920's and early 1930's and elementary particles and structure of nuclei since the mid-1930's are examples.

The situation in mathematics is entirely different. Mathematics falls into a great number of subdivisions, differing from one another widely in character, style, aims, and influence. It shows the very opposite of the extreme concentration of theoretical physics. A good theoretical physicist may today still have a working knowledge of more than half of his subject. I doubt that any mathematician now living has much of a relationship to more than a quarter. "Objectively" given, "important" problems may arise after a subdivision of mathematics has evolved relatively far and if it has bogged down seriously before a difficulty. But even then the mathematician is essentially free to take it or leave it and turn to something else, while an "important" problem in theoretical physics is usually a conflict, a contradiction, which "must" be resolved. The mathematician has a wide variety of fields to which he may turn, and he enjoys a very considerable freedom in what he does with them. To come to the decisive point: I think that it is correct to say that his criteria of selection, and also those of success, are mainly aesthetical. I realize that this assertion is controversial and that it is impossible to "prove" it, or indeed to go very far in substantiating it, without analyzing numerous specific, technical instances. This would again require a highly technical type of discussion, for which this is not the proper occasion. Suffice it to say that the aesthetical character is even more prominent than in the instance I mentioned above in the case of theoretical physics. One expects a mathematical theorem or a mathematical theory not only to describe and to classify in a simple and elegant way numerous and *a priori* disparate special cases. One also expects "elegance" in its "architectural," structural makeup. Ease in stating the problem, great difficulty in getting hold of it and in all attempts at approaching it, then again some very surprising twist by which the approach, or some part of the approach, becomes easy, etc. Also, if the deductions are lengthy or complicated, there should be some simple general principle involved, which "explains" the complications and detours, reduces the apparent arbitrariness to a few simple guiding motivations, etc. These criteria are clearly those of any creative art, and the existence of some underlying empirical, worldly motif in the background—often in a very remote background—overgrown by aestheticizing developments and followed into a multitude of labyrinthine variants, all this is much more akin to the atmosphere of art pure and simple than to that of the empirical sciences.

You will note that I have not even mentioned a comparison of mathematics with the experimental or with the descriptive sciences. Here the differences of method and of the general atmosphere are too obvious.

I think that it is a relatively good approximation to truth—which is much too complicated to allow anything but approximations—that mathematical ideas originate in empirics, although the genealogy is sometimes long and obscure. But, once they are so conceived, the subject begins to live a peculiar life of its own and is better compared to a creative one, governed by almost entirely aesthetical motivations, than to anything else and, in particular, to an empirical science. There is, however, a further point which, I believe, needs stressing. As a mathematical discipline travels far from its empirical source, or still more, if it is a second and third generation only indirectly inspired by ideas coming from "reality," it is beset third with very grave dangers. It becomes more and more purely aestheticizing, more and more purely *l'art pour l'art*. This need not be bad, if the field is surrounded by correlated subjects, which still have closer empirical connections, or if the discipline is under the influence of men with an exceptionally well-developed taste. But there is a grave danger that the subject will develop along the line of least resistance, that the stream, so far from its source, will separate into a multitude of insignificant branches, and that the discipline will become a disorganized mass of details and complexities. In other words, at a great distance from its empirical source, or after much "abstract" inbreeding, a mathematical subject is in danger of degeneration. At the inception the style is usually classical; when it shows signs of becoming baroque, then the danger signal is up. It would be easy to give examples, to trace specific evolutions into the baroque and the very high baroque, but this, again, would be too technical.

In any event, whenever this stage is reached, the only remedy seems to me to be the rejuvenating return to the source: the reinjection of more or less directly empirical ideas. I am convinced that this was a necessary condition to conserve the freshness and the vitality of the subject and that this will remain equally true in the future.

I always thought that the goal of science was to glorify the human spirit.

— Karl Gustav Jacobi (1804–1851)

The Community of Scholars

André Lichnerowicz

André Lichnerowicz (1915–1998) was born in Bourbon l'Archambault a town in the Auvergne located in south central France near Clermond-Ferrand. It is a picturesque town boasting many spas.

One of his ancestors was from Poland—hence his Polish-sounding name. His parents were both teachers; his father was secretary general of the Alliance Française while his mathematician mother was from the École Normale of Sèvres.

In 1939 he wrote a thesis in differential geometry and general relativity theory—subjects that held his interest throughout his professional life and to which he made numerous fundamental contributions. In 1941 he was appointed to a position in Strasbourg but when the city was occupied by the advancing German armies he went to Clermont-Ferrand. A German raid on this town resulted in his capture but he managed to escape the invading forces.

He was a very energetic and stimulating person of wide interests; he could discourse entertainingly on a wide variety of subjects—French history, literature, geography and of course French wines. Together with his Peruvian wife who taught Spanish, they formed a hospitable and stimulating couple.

He was very active in the movement for the reform of mathematics education in France from 1966 till 1973. He had many students, many of whom achieved recognition both at home and abroad. It was said of him that he had an uncanny ability to tailor a thesis topic both to the student's abilities and interests.

As the accompanying article suggests, he had deep interests in the relation of mathematics to the wider world. He also thought deeply about the nature of his subject. In 1987, he wrote as follows: "A mathematician is first of all an artisan learning by throwing himself against his own spirit, a necessary humility. He dreams and is a bit of an artist ... I believe that if my neurophysiologist colleagues took electroencephalograms of mathematicians, they would discover no difference between those of a working mathematician and a composer of music.... Mathematics carries a form of witness of all that the spirit of humans have in common, since mathematics does not depend on a civilization or a culture." And again, in an interview he was confronted with the question: "You show us many beautiful and impressive formulae, many sophisticated constructions and self-consistent theories. But what is the relation of all this to reality?" His answer: "Mathematics is not interested in the nature of things; it puts 'being' in parentheses; this gives it at the same time its power and ambivalence: it is radically non-ontological."

He died in 1998 having received, during his lifetime, numerous honors and great admiration.

Editor's Preface

In this essay, Lichnerowicz expresses his views on what characterizes the community of scholars.

Over the centuries the tension between the community of scholars and the society in which they live and function has waxed and waned. By scholars we mean individuals whose circumstances allow them to spend their lives engaged in speculation and creativity in the humanities, social and natural sciences (including mathematics).

The tension can arise from several sources. The scholar may question the existing political order or may remonstrate against the existing religious hierarchy; the scholar may challenge existing scientific theories or may propose radical departures from the accepted or traditional wisdom. There is however another type of tension: scholars are stigmatized for being parasites—living within a small elite circle contributing nothing to the physical needs of the wider society in which they live. These challenges may and have resulted in various constraints or even persecutions. One of the most notable examples of the latter that comes to mind is Socrates. But other martyrs in various areas are not hard to find: Woolsey, Galileo, etc.

Lichnerowicz sets strict guidelines to which he feels scholars should adhere. Some of these are mirrored in earlier writers. The well-known American philosopher Ralph Waldo Emerson in his essay "The American Scholar" writes that in society, different functions are performed by different individuals but "in this distribution of functions, the scholar is the delegated intellect... he is 'Man Thinking.' In the degenerate state, when he is the victim of society, ... he tends to become the parrot of other men's thinking.... There goes the notion that the scholar should be a recluse... as unfit for any handiwork or public labor." He goes on to say that the "preamble of thought, the transition through which it passes from the unconscious to the conscious is action."

In a somewhat different vein, the great French mathematician Poincaré, in his essay on science and morality, writes as follows: "If science no longer appears as being capable of influencing people's hearts, and as being indifferent to morality, could it not have a detrimental as well as a beneficial effect? And would it not make us lose sight of those things not pertaining to it; the love of truth is without doubt a fine thing but a vain aim if to pursue it we sacrifice goals infinitely more precious such as goodness, compassion, love of neighbor."

The debate is without cease. In contemporary times, a dispute rages in connection with the genome project. The American patent office is permitting scientists to patent the structure of certain molecules, and tempers rise on both sides, some turning away with horror at the thought of giving exclusive rights to scientists who discover the structure of a molecule. Lichnerowicz's essay implies forcefully on which side of the argument he falls.

I have given several lectures in my life. But in almost every case I had the invaluable security of a blackboard and a piece of chalk. The topic was mathematics or physics which are exact sciences and I had only to manipulate these courageous equations, which remained faithfully on the blackboard or were transformed according to the rules of mathematical ballet, but presented hardly any problems of conscience.

It is, believe me, a strange adventure for a mathematician to be constrained to lose the security of his familiar language and be brought to grapple with some of the most serious problems posed by the present condition of the society of mankind. If this brings a certain gravity and some awkwardness, I feel sure that you will pardon this condition. But this constraint which I present with a mixture of joy and despair, is a sign of a constraint infinitely more serious, which presses upon the entire community of scholars.

But what is this scholar whose condition I wish to analyze? Is it he who knows or possesses a certain truth? The question itself is a disparagement of the entire progress of modern science. This science has taught us that these truths are extinct truths whose cadavers are given to children in the form of a manual, for the secondary schools, of approximate truths that are in a state of being rendered obsolete. The realm of the scholar is certainly not that of proprietorship.

For the person of the 18th century, the concept of scholar was clear and described a certain attitude of mind to which we shall return. But in the year 1955, our vocabulary has become confused since it translates the confusion of our minds; we use the words scholar and technician interchangeably and the qualifier "researcher" has recently appeared and for thirty years has experienced an unforeseen fate. I even know a beauty parlor in a street of Paris that calls itself modestly "Institute of Aesthetic Research," a title that threw me into a legitimate perplexity.

There is something sound in this stress on research since the scientific spirit is not a spirit of possession but one of research, of deep investigation. But it is also a source of confusions and these confusions are not at all innocent. To use an extreme example, what then distinguishes a scholar from a technician who is a grand chief engineer, head of a research laboratory of an important electronics firm? Both had been researchers; we have rediscovered—and that is true—that the technical procedures of pure research and those of applied research are indistinguishable. However we perceive a fundamental difference between the attitudes of mind of these two men. Broadly speaking, if you will permit me to speak frankly, the one can find the pinnacle of his career by becoming the CEO of his firm without being a traitor to his vocation while the other not. The one belongs to a highly valued corporation of great use to society while the other is a member of one of the rare spiritual communities that exist in this world—the community of scholars.

It is perhaps this distinction that has been lost sight of in thinking of a researcher, and it is that which I feel compelled to reaffirm with a certain stridency. What then is the mental attitude of a scholar? We can profitably examine this, either by the behavior of contemporary scholars, or by tracing this human attitude across history and across centuries. Since I wish to speak of things of which I am least ignorant, I shall limit the discourse to the exact sciences, and to several implications of these sciences to the domain of the social sciences.

A scholar is a person who participates actively in the scientific adventure, who is a crusader in the scientific enterprise. But this adventure is by its nature a collective one and in order to participate, the scholar should have made certain vows and practiced a certain asceticism; an intellectual asceticism but also a moral one, the two being indissolubly intertwined. If the stress is generally placed on the first, the second is no less important, and is now often called into question for reasons which we shall examine.

This is not the place to describe the scholar at work and analyze the restrictions that have been imposed and that should favor the blossoming of a certain type of imagination and yet assure control and rigor: this necessity of an openness of mind, of a mind ready to gather all that crops up with a deliberate freedom of attention and this mercilessly critical spirit destined, in discarding all speculation, to weave the scientific materials into a circumscribed network which is communicable to anyone who takes the trouble to study it. This absence of respect, in the scientific arena, for all exterior thought, is limited, and at the same time this resolve for total clarity, that sacrifices, without regret, all that which is troublesome or too complicated.

But these disciplines imply and impose moral choices. How can one retain the complete availability of one's mind if we aim, above all, toward applications and profitable technical ones at that? How can he be assured of his own independence, if in his arena, he bows to those in power, or to religious thought or to exterior philosophies? The will to independence, the indifference to applications should be, in a sense I shall make precise, the fundamental elements of the attitude of the spirit of a scholar.

After all, for the scholar, it is more subtle traps to which we more or less all succumb. The scholar has devoted his life to research, but it is very rare that, in the advancing years, the spark continues to glow. In a scientific notebook of Pasteur, we find a marginal note: "On the whole, nothing for two years," and this simple note denotes the anguish, among scholars, of knowing whether the spark has definitely been extinguished or whether the gift of creating science is still available to him. That is why, being a researcher in the real meaning of the term, is not a trade or it is the worst of the trades. In addition to his research, the scholar exercises a true profession, a reassuring profession: he is a professor in a university or administrator in a laboratory. But it then happens that this profession consumes the researcher in him, or inversely the scholar searches for an alibi in his profession.

However that may be, after years of work, he has brought to the total corpus of scientific work a contribution of which none better than he know how limited it is, entwined in all the efforts of a generation and not being evaluated but by the secular work of men of science. This contribution, modest or notable, has on the other hand, gradually ceased to interest him: (this is not difficult since it has already been accomplished) and he does not derive any glory from within himself: the adventure, that is at work, largely overtakes the field of the small personal contributions.

He has dwelt for several years or indeed a lifetime, in the spirit of scientific conquest, he has taken part in the work of the community of scholars, and that is his true honor.

This attitude of mind, of which we have seen a resurgence in our times, has been slowly enhanced over the centuries and it is very likely that it is science that has taught society the essence of intellectual probity.

Greek science started to teach us the rigor of discourse, a rigor which we have little by little reinforced to the limits of contemporary axiomatics, to the point of being able to reason without contradictions on infinite sets and construct mathematics with these axiomatics. But long and painful efforts of modern science were necessary to learn to control certain aspects of what we call the "real numbers," in questioning, with the help of accepted experiments, and in tightening the arguments with new tools of mathematics. The contemporary physical theory is an effort to deduce from mathematics a large class of phenomena, and only the strictest experimental evidence verifies that it is not a vacuous theory; it is a the theory of several imaginary worlds which appear to symbolize the success of scientific aspirations.

But with the idea of experimental hypotheses, with the importance and abundance of designed experiments, there had appeared something new in the progress of science. However, in principal, a mathematician can always verify the proof of another mathematician, and indeed he often engages in this exercise; the physicist, however, uses experimental evidence, that is, the results of many experiments, which he has neither the time nor the resources to redo. He trusts the results of others, he is compelled to have confidence that the members of his community have gone further, and to assume that they have uttered the truth and the entire truth. The integrity of the reports of experiments impose all the restraints and at first prohibit secret things. This integrity is also an effort to spare the need for duplication.

It is in experimental science that what we call the community of scholars appears in its entirety, a community still few in number—France, a great country of science of the 18th C., had only several dozen scholars—but from its first appearance the ideals of science were raised high.

Scientific thought wishes to be totally autonomous, and it avoids secrecy in which it sometimes found refuge in the past. All work which is accomplished should be made public, in order to permit everyone, in complete freedom, to enter the community or to use in a different context, the results that have been acquired. The challenges and secrets of the preceding centuries are regarded as childish and deplorable. Across different countries, universities and academies give benevolent assurance to all, that they have the freedom to engage in research and to disseminate their results. Wars do not, in any way, impede these exchanges and from the 17th C. we see Huygens come to lay siege to the Academy of Sciences of Paris, during a total war between France and Holland, with neither Frenchman, nor Dutchman nor Spaniard finding anything out of the ordinary.

It is true that the applications of this science, beneficent or malevolent, which are produced are almost the exclusive domain of hope. But yet the scientific conscience exerts pressure, and with a candid optimism, judges that in general these applications will be beneficial. It will require a long time to emerge from this optimistic view and we will come to recognize a responsibility in the process of making these matters available to the common people in England and in France, as happened in the first industrial revolution.

However, despite this store of applications, the reaction of the scientific community is formal: the scholar should remain objective, not only in the objects of research but impartial in himself as well. It is to others to assume the great task of applications and of the hard won material advantages for the benefit of all, to perfect the difficulties and subtle procedures of manufacturing. The scholar should have nothing to do with the secret matters, but his lack of interest should not signify that he should be totally unaware of the consequences of his work for mankind.

On the other hand we ignore the ends which can be served by this preoccupation, but as the consequences, in the long run, can only be good; everything is for the best.

Such, sketched with broad strokes, is what we can call the classic ideal of science.

It is this ideal, filled with an old wisdom, which we are led, painfully, to review. In its journey, science has encountered forces. Of the profound transformations which mankind has experienced or permitted, scientific conscience has been one of the most notable victims.

We are all aware that for the past century, our daily universe has been profoundly transformed, has exploded, in all the meanings of this term. This scientific and technical universe, that is ours, an artificial universe which appears more and more like a fabricated universe, an artificial universe which serves both as an incubator and instrument, a universe which can get off track and one that we sense is capable of splintering in a moment of collective insanity. The wave front of human expansion moves so quickly nowadays, and generates such distortions that the responsibility of conditioning society to this world, without ceasing to refashion it, is no longer permitted in the gradual education of new generations. We encounter there, without doubt, some of the reasons for which this universe, admittedly human, appears to us artificial and dangerously alien. We are all endlessly surprised by the events.

This universe by its spirit and by its structures is something else, and it compels each of us, during one's lifetime, to investigations made painful by this new state of equilibrium. We often question inquiries of new economic reactions, as with new schemes of thought, to understand this real movement.

Of the world, still heavy and awkward from the first industrial revolution, a world made of steel castings, and to which steam engines with their crude regulator, conferred some autonomy, we are in the process of making a light world which is knowingly regulated, made of special steels, of aluminum and of magnesium, of glass and plastics, rich with enormous quantities of energy—we are in the process of claiming solar power as well as atomic power—and of commodities subtly controlled by electronic means.

In this world sources of wealth have been profoundly modified and the distortions are more serious than ever. Inhabitants of some countries live, almost without basic resources, from the revenue of their science and their technology imbedded in industries of high precision; others, which historic circumstances have placed outside of the grand stream of scientific expansion, are compelled to produce basic commodities in order to subsist; these countries are called underdeveloped. André Mayer showed they are undernourished, have not benefitted from global medical progress, and experience a population increase out of proportion to their resources, which remain virtually static.

This world which is ours, with its obvious glamor and malevolence, is something we cannot reject. Singing praises of a world in upheaval and brandished with our own curse, these are activities for irresponsible mandarins. We are not permitted to condemn to death those old people whom we have artificially saved, these children, ever more numerous, who have been preserved from epidemics. We must find, first of all, an object in life for the first of these and not abandon them in a desert of a useless old age; we should feed both of these, and we dream of the day when, with a great burst of solar energy, we can manufacture food directly, without passing through the usual agricultural techniques which are too slow. In the forefront of preoccupations of scientific countries are, as we know, photochemistry, and photosynthesis. Already, on the way to a solution, several "algae factories" operate in the world. From the distortions of a scientific world, we compel ourselves to leave more science and a science more conscious of itself.

Another facet of the problem should be noted: science is a tool of forecasting and an entire branch of contemporary science is compelled, with the help of statistical techniques and game theory, to develop precise instruments of forecasting economic phenomena, or more generally, social phenomena, and endeavor to prepare a technique of rational decisions in matters of human conduct. Such a science is, by its nature, a source of power and richness and it is already a reality in several rare applications. But this science which is created and yet still vacillates, cannot be confident except for short-term forecasts—perhaps a few years. Besides, scientific research itself which shows itself as the most formidable factor of instability of our world, opposes every valid prediction: in twenty years it is impossible to predict with success the results of our work.

Science has thus encountered strengths in the material consequences of its results and even in certain of its objectives of research. It has hurriedly tested the weight of its responsibilities before the society of humans. Finally it has been revealed to it that it needed to master its task, even the most classic, and that the ideal of the 18th C. should have been questioned, not for reasons of power, but for the good, for the survival of the science itself.

The primitive experimental initiatives consisted entirely of techniques, and the scholar, helped by a locksmith or mechanic, sufficed for their realization. It was possible to minimize the role of an industry still in limbo. But scientific research now lies in a factory, uses the acquisitions of science not only directly but indirectly beyond its realization, in industrial production. This results in a response on the part of the scholar, to have to leave these applications to others. A large contemporary research laboratory has the dimensions, the usage, the personnel and indeed the methods of a veritable factory, heir to other factories. In certain areas, in nuclear physics for example, a single experiment is an enormous machine grouping around it scholars and technicians by the dozens, and requiring for its full utilization auxiliary laboratories which thirty years ago would have given joy to a physicist.

In large areas, scientific activity passes to the highest level of industrialism, and many scholars, our contemporaries, are overwhelmed and are not able to cope with the immensity of the necessary means.

Science is not a luxurious activity for serious people as it was in the 18th C.; it interests and worries those in power greatly and is led to ask for material means that are no longer those which are suitable for the encouragement of the pleasurable arts, but those, which for a nation, answer to a vital need. The scientist and the financier are compelled to engage in a dialogue, a dialogue filled with ambiguities.

I want to analyze briefly certain of the ambiguities of the dialogue of those in power with the community of scholars. Each has, without doubt, a good conscience and some bad thoughts.

To those in power it would have required, for their private as well as state interests, a singularly elevated point of view to understand instinctively the scientific ideal. As long as science was, if I dare say, an art of agreement, it remains so. But it now concerns serious matters, to win economic battles, briefly to invest considerable sums diverted from the community towards research. It is a matter without doubt infinitely too serious to leave in the hands of scholars.

Those in power sense, rather mistakenly, that in the exercise of their mission, they have no need of scholars, but in fact they need technicians, or if you prefer, researchers in the modern sense of this term. One should make laymen of these clerics. In view of their position, those in power have the choice of decisions and responsibilities; to the technicians employed they must ensure the achievement of the plan, without imposing serious problems. The advance—scientific or technical, it is of little importance—obtained in an area should be preserved and secrecy will conceal it. To the scholars, properly speaking, to those who persist, a certain marginal activity will be permitted; they will also be used to train technicians. Such is, throughout the world, the natural progression of thought of the leaders who, because of their individual experience, cannot but deny the scientific ideal.

The secrecy, in scientific matters, has made a reappearance and we see indeed that this ideal is misunderstood in large matters as well as in small ones. The large ones are too well known for me to return to them, but the small ones can serve as a sign. A director of a private enterprise understands only poorly why a geologist friend, having been given a valuable assignment, refuses all personal compensation; such a politician cannot understand why another scholar refuses to direct a large research organization for fear of becoming a director and no longer a research worker. UNESCO itself has recently tried to study the right of scientific property, the right of a scholar, and has seriously asked whether it concerned the right of creation or the right of discovery. The answer which I was led to give to these questions was the following: There cannot be a personal right of scientific property, but possibly only a collective right which belongs to the community of scholars. The scholar is by definition one who does not place a monetary value on the results of his work but delivers them freely to all. If he wanted to reserve a material gain, he could take royalties. If you are led to recognize a collective right, this choice can only be a moral right, as long as you do not consider, as the opponent, the methods placed at the disposal of scientific research. Most of the scholars consulted replied in the same manner, but this point of view was not satisfactory to the lawyers. But it is the only consistent view for the profession of a scholar.

What I have just said about those in power is, naturally, a bit of a caricature, but the caricature has some truth to it. It should also be noted that those in power are, by their nature, technically incompetent; indeed they are led, in most cases, to follow the suggestions of their technicians and of their experts who are to be found in a struggle without realistic possibilities of compromise. But within many of these technicians, the germ of a scholar is present.

Hence, the scientific community has recently been required to confront new problems concerning its relations with society. It was ill prepared and showed little taste for this issue for which it sensed itself ill equipped. Few scholars were inclined to ponder these problems: scientific tasks appeared to them to be more urgent.

But no one was there to be a substitute for them. It is a curious and saddening fact that the scientific adventure has little interest for the scientific philosophy of our times. Neither Husserl, nor Jaspers, nor Sartre have brought valuable points of view about science. The world of scientific work where we function, remains closed to them and not one of them has undertaken to discern patiently and honestly, the implicit philosophy which is at the heart of scientific thought. But in reality, it is the scholars who should reflect on their own problems. No help will come to them from the outside.

For a long time they restricted themselves, either to exhibit a candid pride of involuntary miracle workers, or taking refuge, if they were malcontents, in some prefabricated political doctrine, or to explain their complete agreement with the authorities: they did not claim to assume any responsibility in this sad and impure history and they did not want to dirty their hands.

Among the capitalists, on the contrary, they tried to justify themselves by the direct utility of research. They explained, at great length, that if free and spontaneous research diminished or disappeared, then directed and applied research would weaken very quickly and would lose most of its power of renewal. This would indeed certainly be the case.

Those in power exclaim: a scholar should not engage in politics. Indeed many scholars take pride in avoiding politics. They claim vaguely their ignorance of the material means of research made available to them. Others participate with head bowed, often extensively, in a political action and find themselves caught in the activities of some faction who use them as authorities.

Many scholars however, find themselves weary of these equally uncomfortable, equally unscientific positions, uncomfortable playing the role of a juggler in a fair. They may resort to a good conscience that emanates from the less noble *pontius pilateism.* Neither the role of the prophetic man nor that of the inoffensive academician suits the scholar. As to the protest of the individual conscience, it is nothing but a childish attitude which imitates *pontius pilateism.*

Taken in the whirlwind of the powers that be, the scientific community has almost lost, in the least honorable manner, its autonomy and, if it is to survive, must face up to it and think about it. This is not a danger if, in the world, the meetings of scholars dealing with these problems increases, and if the journals echo their sentiments. After long discussions, certain of the renowned national or international scientific societies have prohibited their members from participating in those universities whose leaders have not recognized the freedoms of the scholars. This public excommunication has had the effect of exerting remarkably efficacious pressure.

In the heritage of the classical ideal of science, there is an indisputable part without which there would be no living scientific community, but an organization of qualified workers who would quickly disappear. And across winds and seas, our community reaffirms that component that consists of loyalty in discussions, freedom in research and communication, indifference to material gain. But this heritage has become a heavy one since responsibilities toward the society of humans have emerged.

This scientific community is in the process of becoming conscious of itself as a social community which defends, not the material interests of its members, but a voluntary set of morals which preserves the integrity of the scientific conscience. It knows that it should remain alert henceforth in a manner aware of the human consequences of the scientific work. It is compelled to reflect on these consequences and to foresee these with all the resources of the critical imagination of its members.

It should not only teach science but *inform* society of the implications of its results, communicate also its hopes and fears, freed from the spirit of its work. Scientific knowledge has perhaps become the first of the new responsibilities of the scholar, but a knowledge made with the same intellectual probity as the science itself and which does not contribute to a dissociation of the prestige of the science from personal philosophical preferences, a knowledge which enlarges the elements of an authentic scientific culture.

The community of scholars should, accordingly work, in a world that is increasingly technical, toward allowing clear options, to preserve for each person an authentic possibility of control, of choice—a choice which is not a capitulation to the publicity, to the propaganda or the authority that claims competence. It knows that it should increase its influence in the world, to assign ambassadors close to those in power, and make them feel its influence, not by a thirst for power but by the wish to assume, in fact and not formally, that part of responsibility which belongs to it.

These are heavier tasks that the community of scholars should henceforth accomplish, in addition to its essential scientific work. I believe that the community will prove itself worthy of the task.

Descartes commanded the future from his study more than Napoleon from his throne.
— Oliver Wendell Holmes (1809–1894)

History of Mathematics: Why and How

André Weil

André Weil's life (1906–1998) spanned virtually the entire 20th century during which mathematics underwent a dramatic and exciting evolution and transformation. Part of this advancement is due to Weil himself.

He was born in Paris the son of Bernard Weil, a physician, and Selma Reinharz. Bernard belonged to a Jewish family from Alsace (which was part of Germany at the time of his birth) while Selma came from a middle class Austrian-Jewish family that had settled in Rostov-on-Don but eventually moved to Paris. A sister Simone was born in 1909; she had a very sad ending dying in Britain during the war in 1943. The two were close as children and remained so throughout Simone's short life. Trained in philosophy, she is now widely revered for her religious mysticism.

André was precocious and, in addition to his natural gifts for mathematics, he took a great interest in, and had an inherent talent for, languages. Among other things, he had a passion for Sanskrit and the Baghavad Gita, and for Greek poetry. He traveled extensively meeting many mathematicians in his journeys. He was awarded the degree DSc from Paris in 1928. He then spent two years at Aligarh University in India where he absorbed the culture of Buddhism. He was then on the faculty of the University of Strasbourg from 1933 until the outbreak of the Second World War. While in Strasbourg, he was one of the main forces in the establishment of "Bourbaki," a group of mathematicians determined to rewrite the whole of mathematics on a firm logical foundation much as in the style that Euclid achieved for geometry.

At this point in his life, there began an odyssey that did not come to an end until he eventually made his way to the U.S.A. in 1941. The source of the problem was his quasi-pacifism. From his own testimony, he was not a pacifist in the strict sense of the word but he had no intention or desire to participate in the armed services. His avowed reason is that in the First World War, the flower of French youth was decimated and this applied especially to young mathematicians. At the outbreak of WWII, French mathematics had not yet recovered from the catastrophe.

Weil went to Finland where he was hospitably received, and had hoped to make his way to the USA but that was not to be. He was required to return to France where he was imprisoned for not reporting for duty. He was released and eventually made his way, via Marseille, to the USA where he arrived in 1941. He taught at Haverford College and Lehigh University. In 1945 he went to São Paulo University in Brazil where he remained till 1947. He then went to the University of Chicago and the Institute for Advanced Study from which he retired in 1976.

In 1936 he married Eveline, who had recently been divorced from her first husband. She had a son Alain. The marriage was a happy one and the Weils had two daughters, Sylvie and Nicolette. Eveline's death in 1986 was a severe blow. Indeed in his autobiography, Weil describes his life as consisting of the interval from birth to his wife's death.

Weil was a very original thinker and left a large volume of important mathematics but, equally significantly, his creative mind opened new pathways that other mathematicians have successfully pursued.

He is said to have had a strong sense of humor and a sharp wit seasoned with doses of sarcasm directed at what he regarded as infelicitous behavior.

Weil received many honors including the noted Kyoto Prize as well as the Wolf Prize and the Steele Prize of the American Mathematical Society.

Editor's Preface

No one ever questions the pertinence of political history (Henry Ford may have been an exception!) or the history of art or music. Yet historians of mathematics feel the obligation to justify the relevance of this undertaking.

A renowned mathematician and mathematical historian—André Weil—rises to the challenge. Indeed at the time of the 1950 Congress (Cambridge, MA) Weil's book on the historical development of elliptic functions had already appeared and there was soon to appear his history of the theory of numbers entitled *Number Theory: An Approach through History from Hammurapi to Legendre.*

Many accounts of political history consist mainly of a recitation of events and dates and in many traditional history courses in schools, this is the primary content. More thoughtful historians however, endeavor to trace the evolution of political thought through the ages. Consider for example the concept of a democratic form of government. It is a deep concept and not one that would readily come to mind as a method of governing a political entity. Its development went through many stages and the thoughts of many writers, through history, have been distilled and amalgamated to propose a workable system that has functioned successfully in numerous contexts. The search for the origins of these ideas is a valuable adjunct to the ideas themselves.

So it is with the evolution of scientific ideas and, in particular, mathematical ones. A mathematical idea may undergo a metamorphosis over time. The development of the idea may not be "linear," i.e., mathematician A may have contributed a deep mathematical insight that is subsequently shrouded in mist and temporarily forgotten. A historical study may resuscitate the original idea and may shed light on its reincarnation. Weil cites several examples and suggests that many more might be revived if care is taken to delve into their history. Weil also surmises that the personality and life of a creative mathematician may infuse life into that person's creations. Why this should be so seems on the face of it unconvincing but Weil suggests that, for example, knowing that Euler lived in St. Petersburg and that he initially worked for the Russian Navy endows his work with a humanity not otherwise felt.

But there is an additional bonus that comes from exploring the historical genesis of a mathematical idea. This arises as follows: Often in learning mathematics, even at an elementary level, a concept or result is presented in its mature form. By contrast it is, at times, exciting to witness the evolution from conception to fruition. This writer can testify that one of the interesting events experienced was reading one of Euler's memoirs dealing with an important theorem. The original had a clarity of which the reader had been unaware. In a different direction, Weil gives as an example the invention of logarithms. The logarithmic function, in contemporary accounts, is presented as an isomorphism. It is illuminating to see that Napier viewed the logarithm as the relation between the motion of two points moving under different constraints.

What is surprising and noteworthy is that mathematical historians have been active throughout the centuries. The Duc de Montmort, who compiled a history of geometry, wrote to Nicholas Bernoulli as follows: "We have history of painting, of music, of medicine. A history of mathematics would be more interesting and useful.... it could be regarded as a history of the human spirit for it is in this science more than in others that man makes known the excellence of the intelligence that God has given him." Weil cites several historians including a pupil of Aristotle named Eudemus who belonged to the school of the peripatetics. Another of Aristotle's pupils, Theophrastus wrote a history of arithmetic, geometry and astronomy. Histories have appeared through the ages and one of the more notable accounts was given by Jean Étienne Montucla in 1756. In the first half of the 20th century we witness a monumental work by Moritz Cantor and in the late decades, a flood of books on history of mathematics and related topics—a testimony to the growing interest in the origins and historical elucidations of mathematical ideas. And a tribute to Weil's influence and predictions on the significance of historical studies.

My first point will be an obvious one. In contrast with some sciences whose whole history consists of the personal recollections of a few of our contemporaries, mathematics not only has a history but it has a long one, which has been written about at least since Eudemos (a pupil of Aristotle). Thus the question "Why?" is perhaps superfluous, or would be better formulated as "For whom?".

For whom does one write general history? for the educated layman, as Herodotus did? for statesmen and philosophers, as Thucydides? for one's fellow-historians, as is mostly done nowadays? What is the right audience for the art-historian? his colleagues, or the art-loving public, or the artists (who seem to have little use for him)? What about the history of music? Does it concern chiefly music-lovers, or composers, or performing artists, or cultural historians, or is it a wholly independent discipline whose appreciation is confined to its own practitioners? Similar questions have been hotly debated for many years among eminent historians of mathematics, Moritz Cantor, Gustav Eneström, Paul Tannery. Already Leibniz had something to say about it, as about most other topics:

"Its use is not just that History may give everyone his due and that others may look forward to similar praise, but also that the art of discovery be promoted and its method known through illustrious examples." [1]

That mankind should be spurred on by the prospect of eternal fame to ever higher achievements is of course a classical theme, inherited from antiquity; we seem to have become less sensitive to it than our forefathers were, although it has perhaps not quite spent its force. As to the latter part of Leibniz' statement, its purport is clear. He wanted the historian of science to write in the first place for creative or would-be creative scientists. This was the audience he had in mind while writing in retrospect about his "most noble invention" of the calculus.

On the other hand, as Moritz Cantor observed, one may, in dealing with mathematical history, regard it as an auxiliary discipline, meant for providing the true historian with reliable catalogues of mathematical facts, arranged according to times, countries, subject-matters and authors. It is then a portion, and not a very significant one, of the history of techniques and crafts, and it is fair to look upon it entirely from the outside. The historian of the XIXth century needs some knowledge of the progress made by the railway engine; for this he has to depend upon specialists, but he does not care how the engine works, nor about the gigantic intellectual effort that went into the creation of thermodynamics. Similarly, the development of nautical tables and other aids to navigation is of no little importance for the historian of XVIIth century England, but the part taken in it by Newton will provide him at best with a footnote; Newton as keeper of the Mint, or perhaps as the uncle of a great nobleman's mistress, is closer to his interests than Newton the mathematician.

From another point of view, mathematicians may occasionally provide the cultural historian with a kind of "tracer" for investigating the interaction between various cultures. With this we come closer to matters of genuine interest to us mathematicians; but even here our attitudes differ widely from those of professional historians. To them a Roman coin, found somewhere in India, has a definite significance; hardly so a mathematical theory.

This is not to say that a theorem may not have been rediscovered time and again, even in quite different cultural environments. Some power-series expansions seem to have been discovered independently in India, in Japan and in Europe. Methods for the solution of Pell's equation were expounded in India by Bhaskara in the XIIth century, and then again, following a challenge from Fermat, by Wallis and Brouncker in 1657. One can even adduce arguments for the view that similar methods may have been known to the Greeks, perhaps to Archimedes himself; as Tannery suggested, the Indian solution could then be of Greek origin; so far this must remain an idle speculation. Certainly no one would suggest a connection between Bhaskara and our XVIIth century authors.

On the other hand, when quadratic equations, solved algebraically in cuneiform texts, surface again in Euclid, dressed up in geometric garb without any geometric motivation at all, the mathematician will find it appropriate to describe the latter treatment as "geometric algebra" and will be inclined to assume some connection with Babylon, even in the absence of any concrete "historical" evidence. No one asks for documents to testify to the common origin of Greek, Russian and Sanskrit, or objects to their designation as indo-european languages.

Now, leaving the views and wishes of laymen and of specialists of other disciplines, it is time to come back to Leibniz and consider the value of mathematical history, both intrinsically and from our own selfish viewpoint as mathematicians. Deviating only slightly from Leibniz, we may say that its first use for us is to put or to keep before our eyes "illustrious examples" of first-rate mathematical work.

Does that make historians necessary? Perhaps not. Eisenstein fell in love with mathematics at an early age by reading Euler and Lagrange; no historian told him to do so or helped him to read them. But in his days mathematics was progressing at a less hectic pace than now. No doubt a young man can now seek models and inspiration in the work of his contemporaries; but this will soon prove to be a severe limitation. On the other hand, if he wishes to go much further back, he may find himself in need of some guidance; it is the function of the historian, or at any rate of the mathematician with a sense for history, to provide it.

The historian can help in still another way. We all know by experience how much is to be gained through personal acquaintance when we wish to study contemporary work; our meetings and congresses have hardly any other purpose. The life of the great mathematicians of the past may often have been dull and unexciting, or may seem so to the layman; to us their biographies are of no small value in bringing alive the men and their environment as well as their writings. What mathematician would not like to know more about Archimedes than the part he is supposed to have taken in the defense of Syracuse? Would our understanding of Euler's number theory be quite the same if we merely had his publications at our disposal? Is not the story infinitely more interesting when we read about his settling down in Russia, exchanging letters with Goldbach, getting almost accidentally acquainted with the works of Fermat, then, much later in life, starting a correspondence with Lagrange on number theory and elliptic integrals? Should we not be pleased that, through his letters, such a man has come to belong to our close acquaintance?

So far, however, I have merely scratched the surface of my theme. Leibniz recommended the study of "illustrious examples," not just for the sake of esthetic enjoyment, but chiefly so that "the art of discovery be promoted." At this point one has to make clear the distinction, in scientific matters, between tactics and strategy.

By tactics I understand the day-to-day handling of the tools at the disposal of the scientist or scholar at a given moment; this is best learnt from a competent teacher and the study of contemporary work. For the mathematician it may include the use of differential calculus at one time, of homological algebra at another. For the historian of mathematics, tactics have much in common with those of the general historian. He must seek his documentation at its source, or as close to it as practicable; second-hand information is of small value. In some areas of research one must learn to hunt for and read manuscripts; in others one may be content with published texts, but then the question of their reliability or lack of it must always be kept in mind. An indispensable requirement is an adequate knowledge of the language of the sources; it is a basic and sound principle of all historical research that a translation can never replace the original when the latter is available. Luckily the history of Western mathematics after the XVth century seldom requires any linguistic knowledge besides Latin and the modern Western European languages; for many purposes French, German and sometimes English might even be enough.

In contrast with this, strategy means the art of recognizing the main problems, attacking them at their weak points, setting up future lines of advance. Mathematical strategy is concerned with long-range objectives; it requires a deep understanding of broad trends and of the evolution of ideas over long periods. This is almost indistinguishable from what Gustav Eneström used to describe as the main object of mathematical history, viz., "the mathematical ideas, considered historically" [2], or, as Paul Tannery put it, "the filiation of ideas and the concatenation of discoveries." [3] There we have the core of the discipline we are discussing, and it is a fortunate fact that the aspect towards which, according to Eneström and Tannery, the mathematical historian has chiefly to direct his attention is also the one of greatest value for any mathematician who wants to look beyond the everyday practice of his craft.

The conclusion we have reached has little substance, to be sure, unless we agree about what is and what is not a mathematical idea. As to this, the mathematician is hardly inclined to consult outsiders. In the words of Housman (when asked to define poetry), he may not be able to define what is a mathematical idea, but he likes to think that when he smells one he knows it. He is not likely to see one, for instance, in Aristotle's speculations about the infinite, nor in those of a number of medieval thinkers on the same subject, even though some of them were rather more interested in mathematics than Aristotle ever was; the infinite became a mathematical idea after Cantor defined equipotent sets and proved some theorems about them. The views of Greek philosophers about the infinite may be of great interest as such; but are we really to believe that they had great influence on the work of Greek mathematicians? Because of them, we are told, Euclid had to refrain from saying that there are infinitely many primes, and had to express that fact differently. How is it then that, a few pages later, he stated that "there exist infinitely many lines" incommensurable with a given one? Some universities have established chairs for "the history and philosophy of mathematics": it is hard for me to imagine what those two subjects can have in common.

Not so clear-cut is the question where "common notions" (to use Euclid's phrase) end and where mathematics begins. The formula for the sum of the first n integers, closely related as it is to the "Pythagorean" concept of triangular numbers, surely deserves to be called a mathematical idea; but what should we say about elementary commercial arithmetic, as it appears in ever so many textbooks from antiquity down to Euler's potboiler on the same subject? The concept of a regular icosahedron belongs distinctly to mathematics; shall we say the same about the concept of a cube, that of a rectangle, or that of a circle (which is perhaps not to be separated from the invention of the wheel)? Here we have a twilight zone between cultural and mathematical history;

it does not matter much where one draws the borderline. All the mathematician can say is that his interest tends to falter, the nearer he comes to crossing it.

However that may be, once we have agreed that mathematical ideas are the true object of mathematical history, some useful consequences can be drawn; one has been formulated by Tannery as follows (*loc. cit.,* (footnote 3), p. 164). There is no doubt at all, he says, that a scientist can possess or acquire all the qualities needed to do excellent work on the history of his science; the greater his talent as a scientist, the better his historical work is likely to be. As examples, he mentions Châsles for geometry; also Laplace for astronomy, Berthelot for chemistry; perhaps he was also thinking of his friend Zeuthen. He might well have quoted Jacobi, if Jacobi had lived to publish his historical work. [4]

But examples are hardly necessary. Indeed it is obvious that the ability to recognize mathematical ideas in obscure or inchoate form, and to trace them under the many disguises which they are apt to assume before coming out in full daylight, is most likely to be coupled with a better than average mathematical talent. More than that, it is an essential component of such talent, since in large part the art of discovery consists in getting a firm grasp on the vague ideas which are "in the air," some of them flying all around us, some (to quote Plato) floating around in our own minds.

How much mathematical knowledge should one possess in order to deal with mathematical history? According to some, little more is required than what was known to the authors one plans to write about; [5] some go so far as to say that the less one knows, the better one is prepared to read those authors with an open mind and avoid anachronisms. Actually the opposite is true. An understanding in depth of the mathematics of any given period is hardly ever to be achieved without knowledge extending far beyond its ostensible subject matter. More often than not, what makes it interesting is precisely the early occurrence of concepts and methods destined to emerge only later into the conscious mind of mathematicians; the historian's task is to disengage them and trace their influence or lack of influence on subsequent developments. Anachronism consists in attributing to an author such conscious knowledge as he never possessed; there is a vast difference between recognizing Archimedes as a forerunner of integral and differential calculus, whose influence on the founders of the calculus can hardly be overestimated, and fancying to see in him, as has sometimes been done, an early practitioner of the calculus. On the other hand, there is no anachronism in seeing in Desargues the founder of the projective geometry of conic sections; but the historian has to point out that his work, and Pascal's, soon fell into the deepest oblivion, from which it could only be rescued after Poncelet and Chasles had independently rediscovered the whole subject.

Similarly, consider the following assertion: logarithms establish an isomorphism between the multiplicative semigroup of numbers between 0 and 1 and the additive semigroup of positive real numbers. This could have made no sense until comparatively recently. If, however, we leave the words aside and look at the facts behind that statement, there is no doubt that they were well understood by Neper [Napier] when he invented logarithms, except that his concept of real numbers was not as clear as ours; this is why he had to appeal to kinematic concepts in order to clarify his meaning, just as Archimedes had done, for rather similar reasons, in his definition of the spiral. [6] Let us go further back; the fact that the theory of the ratios of magnitudes and of the ratios of integers, as developed by Euclid in Books V and VII of his Elements, is to be regarded as an early chapter of group-theory is put beyond doubt by the phrase "double ratio" used by him for what we call the square of a ratio. Historically it is quite plausible that musical theory supplied the original motivation for the Greek theory of the group of ratios of integers, in sharp contrast with the purely additive treatment of fractions in Egypt; if so, we have there an early example of the mutual interaction between pure and applied mathematics. Anyway, it is impossible for us to analyze properly the contents of Books V and VII of Euclid without the concept of group and even that of groups with operators, since the ratios of magnitudes are treated as a multiplicative group operating on the additive group of the magnitudes themselves. [7] Once that point of view is adopted, those books of Euclid lose their mysterious character, and it becomes easy to follow the line which leads directly from them to Oresme and Chuquet, then to Neper and logarithms. In doing so, we are of course not attributing the group concept to any of these authors; no more should one attribute it to Lagrange, even when he was doing what we now call Galois theory. On the other hand, while Gauss had not the word, he certainly had the clear concept of a finite commutative group, and had been well prepared for it by his study of Euler's number-theory.

Let me quote a few more examples. Fermat's statements indicate that he was in possession of the theory of the quadratic forms $X^2 + nY^2$ for

$n = 1, 2, 3$, using proofs by "infinite descent." He did not record those proofs; but eventually Euler developed that theory, also using infinite descent, so that we may assume that Fermat's proofs did not differ much from Euler's. Why does infinite descent succeed in those cases? This is easily explained by the historian who knows that the corresponding quadratic fields have an Euclidean algorithm; the latter, transcribed into the language and notations of Fermat and Euler, gives precisely their proofs by infinite descent, just as Hurwitz' proof for the arithmetic of quaternions, similarly transcribed, gives Euler's proof (which possibly was also Fermat's) for the representation of integers by sums of 4 squares.

Take again Leibniz' notation $\int ydx$ in the calculus. He insisted repeatedly on its invariant character, first in his correspondence with Tschirnhaus (who showed no understanding for it), then in the *Acta Eruditorum* of 1686; he even had a word for it ("*universalitas*"). Historians have hotly disputed when, or whether, Leibniz discovered the comparatively less important result that, in some textbooks, goes by the name of "the fundamental theorem of the calculus." But the importance of Leibniz' discovery of the invariance of the notation ydx could hardly have been properly appreciated before Elie Cartan introduced the calculus of exterior differential forms and showed the invariance of the notation $ydx_1 \ldots dx_m$, not only under changes of the independent variables (or of local coordinates), but even under "pull-back." [8]

Consider now the debate that arose between Descartes and Fermat about tangents. Descartes, having decided, once and for all, that only algebraic curves were a fit subject for geometers, invented a method for finding their tangents, based upon the idea that a variable curve, intersecting a given one C at a point P, becomes tangent to C at P when the equation for their intersections acquires a double root corresponding to P. Soon Fermat, having found the tangent to the cycloid by an infinitesimal method, challenged Descartes to do the same by his own method. Of course he could not do that; being the man he was, he found the answer (*Oeuvres,* II, p. 308), gave a proof for it ("quite short and quite simple") by using the instantaneous center of rotation which he invented for the occasion, and added that he could have supplied another proof "more to his taste and more geometrical" which he omitted "to save himself the trouble of writing it out"; anyway, he said, "such lines are mechanical" and he had excluded them from geometry. This, of course, was the point that Fermat was trying to make; he knew, as well as

Descartes, what an algebraic curve was, but to restrict geometry to those curves was quite alien to his way of thinking and to that of most geometers in the XVIIth century.

Gaining insight into a great mathematician's character and into his weaknesses is an innocent pleasure that even serious historians need not deny themselves. But what else can one conclude from that episode? Very little, as long as the distinction between differential and algebraic geometry has not been clarified. Fermat's method belonged to the former; it depended upon the first terms of a local power series expansion; it provided the starting point for all subsequent developments in differential geometry and differential calculus. On the other hand, Descartes' method belongs to algebraic geometry, but, being restricted to it, it remained a curiosity until the need arose for methods valid over quite arbitrary ground-fields. Thus the point at issue could not be and was not properly perceived until abstract algebraic geometry gave it its full meaning.

There is still another reason why the craft of mathematical history can best be practiced by those of us who are or have been active mathematicians or at least who are in close contact with active mathematicians; there are various types of misunderstandings of not infrequent occurrence from which our own experience can help preserve us. We know only too well, for instance, that one should not invariably assume a mathematician to be fully aware of the work of his predecessors, even when he includes it among his references; which one of us has read all the books he has listed in the bibliographies of his own writings? We know that mathematicians are seldom influenced in their work by philosophical considerations, even when they profess to take them seriously; we know that they have their own way of dealing with foundational matters by an alternation between possibly reckless disregard and the most painful critical attention. Above all, we have learnt the difference between original thinking and the kind of routine reasoning which a mathematician often feels he has to spin out for the record in order to satisfy his peers, or perhaps only to satisfy himself. A tediously laborious proof may be a sign that the writer has been less than felicitous in expressing himself; but more often than not, as we know, it indicates that he has been laboring under limitations which prevented him from translating directly into words or formulas some very simple ideas. Innumerable instances can be given of this, ranging from Greek geometry (which perhaps was at last suffocated by such limitations) down to the so-called epsilontic and down to Nicolas Bourbaki, who even once considered using a special sign in the margin to warn the reader about proofs of that kind. One important task of the serious historian of mathematics, and sometimes one of the hardest, is precisely to sift such routine from what is truly new in the work of the great mathematicians of the past.

Of course mathematical talent and mathematical experience are not enough for qualifying as a mathematical historian. To quote Tannery again (*loc. cit.* (footnote 3), p. 165), "what is needed above all is a taste for history; one has to develop a historical sense." In other words, a quality of intellectual sympathy is required, embracing past epochs as well as our own. Even quite distinguished mathematicians may lack it altogether; each one of us could perhaps name a few who resolutely refuse to be acquainted with any work other than their own. It is also necessary not to yield to the temptation (a natural one to the mathematician) of concentrating upon the greatest among past mathematicians and neglecting work of only subsidiary value. Even from the point of view of esthetic enjoyment one stands to lose a great deal by such an attitude, as every art-lover knows; historically it can be fatal, since genius seldom thrives in the absence of a suitable environment, and some familiarity with the latter is an essential prerequisite for a proper understanding and appreciation of the former. Even the textbooks in use at every stage of mathematical development should be carefully examined in order to find out, whenever possible, what was and what was not common knowledge at a given time.

Notations, too, have their value. Even when they are seemingly of no importance, they may provide useful pointers for the historian; for instance, when he finds that for many years, and even now, the letter K has been used to denote fields, and German letters to denote ideals, it is part of his task to explain why. On the other hand, it has often happened that notations have been inseparable from major theoretical advances. Such was the case with the slow development of the algebraic notation, finally brought to completion at the hands of Viète and Descartes. Such was the case again with the highly individual creation of the notations for the calculus by Leibniz (perhaps the greatest master of symbolic language that ever was); as we have seen, they embodied Leibniz' discoveries so successfully that later historians, deceived by the simplicity of the notation, have failed to notice some of the discoveries.

Thus the historian has his own tasks, even though they overlap those of the mathematician and may at times coincide with them. Thus, in the

XVIIth century, it happened that some of the best mathematicians, in the absence of immediate predecessors in any field of mathematics except algebra, had much work to do which in our view would fall to the lot of the historian—editing, publishing, reconstructing the work of the Greeks, of Archimedes, Apollonios, Pappos, Diophantos. Even now the historian and the mathematician will not infrequently find themselves on common ground when studying the production of the XIXth and XXth centuries, not to mention anything of more ancient vintage. From my own experience I can testify about the value of suggestions found in Gauss and in Eisenstein. Kummer's congruences for Bernoulli numbers, after being regarded as little more than a curiosity for many years, have found a new life in the theory of p-adic L-functions, and Fermat's ideas on the use of the infinite descent in the study of Diophantine equations of genus I have proved their worth in contemporary work on the same subject.

What, then, separates the historian from the mathematician when both are studying the work of the past? Partly, no doubt, their techniques, or, as I proposed to put it, their tactics; but chiefly, perhaps, their attitudes and motivations. The historian tends to direct his attention to a more distant past and to a greater variety of cultures; in such studies, the mathematician may find little profit other than the esthetic satisfaction to be derived from them and the pleasures of vicarious discovery. The mathematician tends to do his reading with a purpose, or at least with the hope that some fruitful suggestion will emerge from it. Here we may quote the words of Jacobi in his younger days about a book he had just been reading: "Until now," he said, "whenever I have studied a work of some value, it has stimulated me to original thoughts; this time I have come out quite empty-handed." As noted by Dirichlet, from whom I have borrowed this quotation, it is ironical that the book in question was no other than Legendre's *Exercices de calcul intégral*, containing work on elliptic integrals which soon was to provide the inspiration for Jacobi's greatest discoveries; but those words are typical. The mathematician does his reading mostly in order to be stimulated to original (or, I may add, sometimes not so original) thoughts; there is no unfairness, I think, in saying that his purpose is more directly utilitarian than the historian's. Nevertheless, the essential business of both is to deal with mathematical ideas, those of the past, those of the present, and, when they can, those of the future. Both can find invaluable training and enlightenment in each other's work. Thus my original question "Why mathematical history?" finally reduces itself to the question "Why mathematics?", which fortunately I do not feel called upon to answer.

Endnotes

[1] *"Utilissimum est cognosci veras in inventionum memorabilium origines, praesertim earum, quae non casu, sed vi meditandi innotuere. Id enim non eo tantum prodest, ut Historia literaria suum cuique tribuat et alii ad pares laudes invitentur, sed etiam ut augeatur ars inveniendi, cognita methodo illustribus exemplis. Inter nobiliora hujus temporis inventa habetur novum Analyseos Mathematicae genus, Calculi differentialis nomine notum...* (*Math. Schr.*, ed. C. I. Gerhardt, t. V, p. 392).

[2] *Die mathematischen Ideen in historischer Behandlung* (*Bibl. Math.* 2 (1901), p. 1)

[3] *La filiation des idées et l'enchaînement des découvertes* (P. Tannery, *Oeuvres,* vol. X, p. 166)

[4] Jacobi, as a student, had hesitated between classical philology and mathematics; he always retained a deep interest in Greek mathematics and mathematical history; extracts from his writings on this subject have been published by Koenigsberger in his biography of Jacobi (incidentally, a good model for a mathematically oriented biography of a great mathematician): see L. Koenigsberger, *Carl Gustav Jacob Jacobi,* Teubner, 1904, pp. 385–395 and 413–414.

[5] Such seems to have been Loria's view: "Per comprendere e giudicare gli scritti appartenenti alle età passate, basta di essere esperto in quelle parti delle scienze che trattano dei numeri e delle figure e che si considerano attualmente come parte della cultura generale dell'uomo civile" (G. Loria, *Guida allo Studio della Storia delle Matematiche,* U. Hoepli, Milano, 1946, p. 271).

[6] Cf. N. Bourbaki, *Eléments d'histoire des mathématiques*, Hermann, 1966, pp. 167–168 and 174; that collection of historical essays, extracted from the same author's *Eléments de mathématique* under a misleading title, will be quoted henceforth as NB.

[7] Whether or not Euclid believed the group of ratios of magnitudes to be independent of the kind of magnitudes under study is still a moot point; cf. O. Becker, *Quellen u. Studien* 2 (1933), pp. 369–387.

[8] Cf. NB, p. 208, and A.Weil, *Bull. Amer. Math. Soc.* 81 (1975), 683.

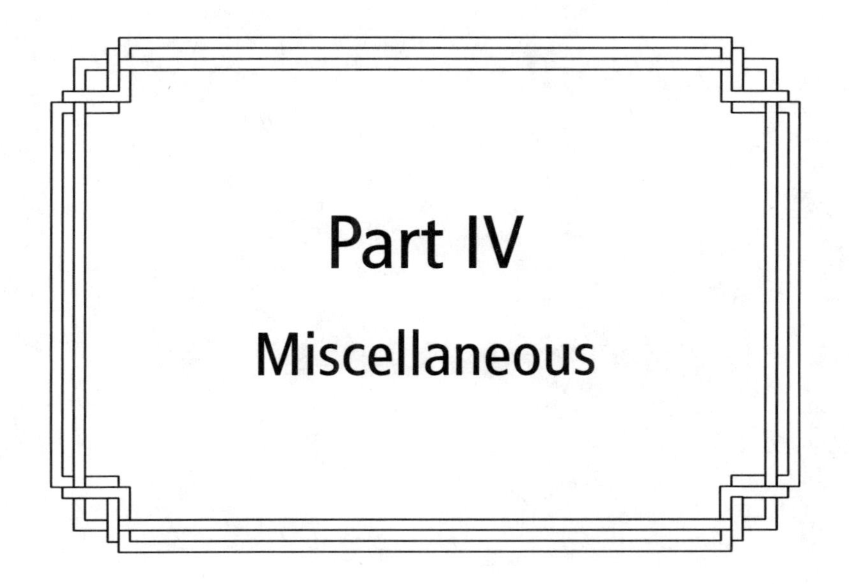

Part IV

Miscellaneous

There are no sects in geometry.

—Voltaire (1694–1778)

Does God Exist?

Paul Pierre Lévy

Paul Lévy (1886–1971) was born in Paris into a family having a long tradition of mathematical activity. His grandfather was a professor of mathematics and his father Lucien Lévy was an examiner at the École Polytechnique. His daughter became a mathematician and married the noted mathematician, Laurent Schwartz.

Lévy exhibited academic aptitude at an early age, excelling not only in mathematics but in languages and science as well. His precocity showed itself when at the age of 19 he published his first paper. After serving a year of military service, he obtained a doctoral degree for a thesis in functional analysis. He had studied with the prominent mathematicians Darboux, Hadamard, Humbert, and Picard. In 1914, after the outbreak of the First World War, he was mobilized into the artillery where he used his mathematical skills to devise a strategy for defense against warplanes. He was appointed professor at the École Polytechnique in 1920, where he remained until his retirement in 1959.

Lévy's main work was in the realm of probability theory and stochastic processes. The history of this discipline goes far back in the history of mathematics but though many results had been proved, probability theory remained in the backwater of mathematics despite the prominence of mathematicians who contributed to it. This neglect may be accounted for by the fact that probability theory was often associated with gambling and games of chance. It remained for Lévy to place the subject on a firm foundation—a foundation as firm as that of other mathematical disciplines. It should be added that the Russian mathematicians Kolmogorov and Khinchin successfully embarked on similar missions.

Lévy was endowed with a powerful mathematical intuition. He was very prolific and was appreciated abroad before recognition came to him at home in France. He was eventually honored by being appointed to the Académie des Sciences in 1964; the London Mathematical Society had elected him as an honorary member in 1963. Since he served as professor of analysis at the École Polytechnique for most of his professional career, he had few research students.

He is described as a very kind and modest man who had wide-ranging interests. In particular, he was interested in metaphysics and in the nature of mathematics. He was critical of philosophical speculation that was not grounded in rational thought. The essay to follow is a critique of what he regarded as essentially unproved speculation. As a mathematician, he is philosophically in the camp of "idealists"—mathematical objects for Lévy had a definite existence.

Editor's Preface

In this essay, Lévy writes on a challenging question that has been a theme of theologians and philosophers for millennia. To be sure, he is more concerned with the God of the Judeo-Christian tradition than with the polytheism and pantheism of other cultures.

We should begin by reviewing attributes of what, in traditional western cultures, is meant by God. These include, but are not limited to: creator of the universe, creator of mankind, prime mover, perfect being, infinitely wise, infinitely good, omniscient, omnipotent, compassionate, merciful, having all the attributes given in the Bible, etc.

There are historically several proofs that such a being exists. We give a very brief summary.

(1) Ontological proof. This is also called the a priori argument. It was first proposed by the theologian Anselm but was taken up by Descartes as well. It could not be simpler: existence is clearly one of the properties of perfection, Q.E.D.
(2) Cosmological argument. This is due to St. Thomas Aquinas. Every action has a cause; there must therefore be a first cause.

(3) Teleological argument or argument by purpose or design. Many natural objects and processes appear to have a purpose. Thus, they are the product of an omnipotent supernatural intelligence.
(4) There are other arguments—the moral argument, the ethical argument, the esthetical argument.

Needless to say volumes have been written refuting or defending these arguments. Pascal, for example, conceded that a belief in God's existence cannot be supported by argument or evidence but he maintained that religious belief is nevertheless required. Moreover, Kant rejected all the above arguments but based his beliefs on the existence and perception of the "Moral Law."

More contemporary theologians and philosophers have examined the question: as an instance, we cite the Swiss theologian Hans Küng who devoted a monumental treatise to the subject.

We come now to Lévy. He brings quasi-mathematical methodology to bear on the problem. He insists that in any discussion we should use logic, right judgment, reasoning, credulity and skepticism. We are not infallible, he writes, but we must reject the idea that nature deliberately deceives us; there is no "malevolent demon." In the realm of metaphysics, nothing is truly proved but our discernment convinces us of the truth of many things.

Lévy acknowledges that at one point in his life he came to the realization that he was, in fact, an atheist. This realization was based, in part, on the observation that if God existed, He would not have allowed all the crimes of war and persecution to be carried out in His name. Moreover, he says, it is easy to account for the creation of a religion: natural phenomena, death in particular, are more easily explained in the context of a religious framework. The peril to this view, he says, is based on what he regards as the gullibility of people accepting doctrine and dogma without questioning. Religious education can be persuasive if it is not carefully scrutinized by the communicant. God is a creation of the soul, he says, and belief simply replaces one mystery by another.

Finally, he is convinced that scientific progress will ultimately lay to rest any belief in God. But creation is filled with wonderment and Lévy is convinced that the world in which we live cannot be the result of chance alone. He questions the validity of an evolution that is based only on chance mutations alone. Here he appeals to a probabilistic observation: while he does not give the explicit calculations, he states that using chance mutations alone, the world would have to be at least ten billion years old. He is convinced that there is a goal to the entire universe but this thought, he says, simply replaces one mystery by another. The point of view that the universe has a component of "intelligent design" is being hotly debated in the ranks of biologists, theologians, and cosmologists. It is not likely that the matter will be settled in the near or distant future. This article is reprinted from Lévy's book entitled *Quelques Aspects de la Pensée d'un Mathématicien.*

Let us begin by examining methods which allow us to have an opinion about this question. I have spoken several times about logic and discernment of reasoning, of experience, of credulity, of skepticism, briefly of all the reasons which determine beliefs and opinions of individuals. The time has come to discuss the value of these different arguments.

I wish first to restore rational opinion. That is relatively simple if we focus our attention on action. I have already called attention to this in speaking of Descartes; we are not given time to reflect on what is reasonable. We often act on instinct without any thought; but rational judgement cannot fail to intervene in everyday life, as well as in our opinions, notably political issues. But should it intervene in scientific research and in reflections on metaphysical problems? This question is more sensitive.

We cannot deny that rational opinion is very relative. That of John is not that of James, and I am the first to recognize that if, in a discussion, my opponent invokes his opinion, it will not be a convincing argument for me. It should be noted that opinion is educable; recollection of errors contributes to this. The opinion of primitive people is not that of an urbane person, nor is it that of a scholar. We have seen, in past times, debates between physicists and geologists concerning the age of our planet; we cannot say that the physicists at that time had lacked perception. They could not surmise that all matter was a formidable reservoir of energy. They inferred it only after the discovery of radioactivity; I am tempted to say that those who, after the discovery of radium, continued to doubt [that all matter is a reservoir of energy] were lacking in discernment.

Let us speak of scientific research in general. In every branch of science, we find ourselves at each moment confronted by numerous problems; it is with our perceptiveness that we choose those that we seek to solve, taking into account their interest and the means available to us. Often too, the method of research depends upon the idea we had a priori of the solution. A mathematician would proceed in different ways according as he wishes to prove a general theorem or, believing it not true in general, he seeks a counterexample. In experimental sciences, we are guided by some assumption, and it is insight that seeks to choose the best hypothesis.

Evidently we achieve unquestioned progress when the result predicted by perception (or by intuition which is a superior form of perception) could be proved by reasoning or by experiments which anyone can carry out. But could we not, in the experimental sciences, question the word "prove?" We could have successfully made an experiment a hundred times, and believed that the experiment will always succeed, and yet be mistaken, either because there is a high probability of success or we have noticed a circumstance favorable to success, a circumstance that suddenly ceases. In the experimental sciences, there is never a true proof in the mathematical sense of the word. It is perception that leads us to consider certain laws as having been proved and to rely on these for further research. Only logic and pure mathematics lead to a greater level of certainty. But what should we say to someone who questions their rules, if not that it is our good judgement that obligates us to accept them. Let us acknowledge the great role played by our judgement and let us rejoice when it guides us in regions where we find nothing that is truly certain. It goes without saying that we should not imitate those who say any old thing in the name of judgement. A "reasoning by judgement" ought to be weighed carefully; it can be questioned. This is not without value. To sum up, I shall say that judgement is to man what instinct is to animals. There is no way around it.

I shall be more concise in dealing with reasoning and experiment, which we often find faulty, especially when we study the foundations of the calculus of probabilities. These are two methods of finding truth and the truth is that these should never contradict one another. But man is not infallible. He can make a mistake in his calculations or in his reasoning; he can also, in an experiment, neglect an important condition or he could misinterpret the result. I do not want to measure the judgement of a physicist who one day announces in front of the Philosophical Society: "We win every time we engage in an experiment." No! If we find a contradiction, we should continue to examine with care both the reasoning and the conditions of the experiment until we find the error that has been committed. We should be very audacious in affirming a priori, that it is the reasoning that is at fault.

I shall approach this remark using an idea of Bergson. He observed that nature is utilitarian: she gives to living things qualities that are needed to live in the environment in which they might find themselves. Nature has given us an intelligence that allows us to understand certain aspects of the exterior world, and we determine that that will enable us to function in this world in a singularly efficacious manner. She has not wanted to deceive us—without doubt, I repeat, we are not infallible—but we must reject the idea that the nature of our reasoning condemns us to commit errors. The malevolent demon, imagined by Descartes, that will entice us into error, does not exist, and I do not need to believe in God to be certain of that. It is only necessary to observe that the idea of the world which we hold directly from nature is limited to a determined environment and to a prescribed degree. That which takes place at a microscopic level or an astronomic level is not directly accessible to our senses. But there is no reason to doubt that we can reason validly about the facts that electron or proton microscopes have shown, and arrive at a correct idea of what the laws of nature are at the level thus attained. I even believe that the physicist, who is sufficiently experienced in thinking about these laws, can be led to apply them as naturally as we apply the laws of gravity to fire a ball to a given distance. It is examples of this nature that came to mind when I said above that judgement can be cultivated and developed. Here on the other hand, I speak of intuition rather than judgement. The physicist in the study of elementary particles of matter should arrive at a special intuition.

Let us return to metaphysics. The reader will agree without difficulty that it is a domain where, outside our existence, nothing is truly proved, but arguments of rational judgement can persuade us of certain things, notably the existence of the exterior world. I consider also as virtually certain, not only that the soul is closely linked to the body, but that it cannot exist without the body, and that it ceases to exist when the conditions sustaining life cease, as the flame goes out when the fuel is exhausted. We can relight the flame; it would not be the same. Moreover, I no longer even conceive that there is a problem with the immortality of the soul. That does not prevent me from considering, as an incomprehensible mystery, the fact that a bit of grey matter can think.

Is this the miracle that proves the existence of God? In any case it is not the second miracle of which I have spoken, which was a breach of the laws of nature decided by God. It is the third miracle, which encompasses all the mystery of the world, a mystery that we cannot hope to clarify completely. Does it prove that God exists? I can do no other than to recount the evolution of my ideas on this matter.

I take up the account of this evolution at the end of the summer of 1902. I had arrived at the conclusion that we cannot truly prove the existence of God, and no longer thought about this problem. I cannot say at which moment I realized that I had become a total atheist; that was no doubt between 1904 and 1908. I cannot even say with certainty what reasons had led me to the conviction that God does not exist. I only think it probable that there were two distinct reasons.

The first is quite simple. If I tried to imagine that He existed, I made of Him so grand an idea that I was convinced that He would not have permitted all the crimes which, over the centuries, were committed in His name. These crimes were certainly committed; it is an incontestable fact. Hence He does not exist.

The other is based on the idea that, besides that which the crudest judgement teaches us, we should not take into account the opinion of the masses. They are credulous; they believe what we tell them. More exactly, there are affinities which lead to the fact that Peter believes John but does not believe James. James might well have explained his reasons which justified his assertions but it is John that Peter believes. Moreover each religion, as with each political party, creates a vocabulary that is imposed upon its members and prevents them from thinking. I could give many examples. I shall limit myself to a person in my family, a convinced communist, who in 1956 had a moment of doubt upon learning of the crimes committed by the Russians in Budapest. The person confronted a "comrade" who said: "But the victims were counterrevolutionaries; they deserve no pity whatsoever." This word "counterrevolutionary" sufficed; the individual's compassion ceased and he no longer had pity for the poor Hungarians and no longer dared blame the Russians. And yet it is a person who devotes his life to humanitarian works.

It is thus very easy to explain religions by the gullibility of the masses, by their tendency to accept a word that appears to them to be an explanation. Lightning startled the ancient Greeks: they were told "it is Zeus who launches it" and they were satisfied. Death frightened the Christians; they were told "it is the punishment for the sin of Adam" and they believed it. Furthermore, different religions contradict one another, and the wisest course of action is not to believe any of them. On the other hand, we can believe in God without adopting a particular religion. There are "souls" comforted by this idea. But for the rationalist, which I am, this God is required by the soul and hence is invented by it. This idea is the most plausible to explain the fact that belief in God is so widespread; for, I repeat, this belief does not help at all in explaining the mystery of the world. We do not explain a mystery by imagining a greater one. I shall shortly return to this point.

I want to express another idea. As far as I am concerned, it is the scientific spirit which has destroyed belief in God. I assumed that this spirit would destroy it in all humans. I would also have believed, a priori, that the two world wars would have negated religions; one would have thought that God, if he exists, would not have permitted them. In the presence of these catastrophes, we would have listened to the voices that chanted human misery rather than that of reason.

And yet I still believe that science will end by denying religions, not as I had first believed but in truly proving that God does not exist. But the development of the scientific spirit will lead humans more and more to reexamine the fundamentals of their beliefs, and not believe blindly that which they were taught in their childhood. We must "reconsider" religions and it seems to me to be fatal that they do not withstand well this perpetual examination. How long will their decadence last? Will it require several generations or several centuries? I do not know. But I believe that in several millennia, we shall no longer consider the gothic cathedrals as vestiges of a vanished religion, and nothing will replace this persuasion.

It remains for me to return to the idea of those who believe in God without adopting a particular religion. Who has made this marvelous world in which we live, if it is not God? In other words, that which I called the third miracle, the only true miracle, that of the existence of laws which have allowed life and this splendid evolution, which has for the present ended in man. Does this prove the existence of God? I said above that "God explains nothing." I understand by that that if He created the world, I do not know how He did it. Thus, for me nothing is explained. The mystery remains in its entirety, and another is added: what is God? But that does not prove that we must adopt the hypothesis of the existence of the world without God. Adopting one or the other hypothesis still leaves us in a state of mystery.

I am sometimes told: "you attest to the existence of a mystery. This is the mystery that we call God." We can evidently change the meanings of the words. But for me "believing in God" means believing in the existence of a being who thinks, who observes us, who reads our thoughts, who judges us, who undoubtedly loves us, who finally is responsive to our prayers and whom we can love. In that sense I do not believe in God. But there has been an evolution in my thinking on the subject of the mystery of the world.

When I was twenty, my maternal grandfather sometimes said to me in springtime: "what a proof of God's existence is the sap which rises in the trees." I did not reply but I thought prosaically: "not at all: it is the capillary force which causes the sap to rise." Two years later, walking in the Luxemburg gardens with a Catholic friend who was a believer, I recounted this conversation and my reaction. He said to me: "But who made capillarity?" This reply was wise, but I was not ready to understand it. The miracle of the world seemed to me to be entirely natural.

And yet, little by little, I discovered that science never does anything but explain one mystery by another. Science never succeeds in discovering a definitive explanation. Thus Newton discovered the law of universal gravitation, which explains the relative movements of the planets and their satellites; but why do the heavenly bodies attract one another. Einstein, in discovering relativity theory, throws new light on these questions. He forces us to reconsider the notions of space and time, which had seemed to be clear. But the mystery did not vanish.

One could make analogous remarks relative to the mystery of life. The creation of a new being by the sexual act is certainly something mysterious. Science achieved significant progress when it discovered that this fruitfulness resulted from the encounter of ovum and sperm; and science made other finds in discovering chromosomes and genes contained within these. But nothing was explained. That this encounter launches a process which leads to a new being, whose genes determine the essential characteristics, is this not a new mystery? Nowadays we are told of the elements that carry out this marvelous creation of a living being. But where does this power lie?

Thus whenever we resolve a problem, we pose a new one. In explaining one mystery, we pose another one. We never arrive at a definite explanation.

What position can the philosopher adopt in the presence of these inexplicable mysteries, and certainly inexplicable for biological sciences? There are no other elements of evaluation except those given by these sciences, and the philosopher has no alternative but to apply judgement to their consideration. Among the elements, except those given by genetics, the most valuable are those given by the theory of evolution. However, a consideration of these elements leads to a strange contradiction. If we deny finality, if we only agree that only chance could produce mutations and that natural selection has assured the survival of those species best equipped to survive the struggle for life, we cannot avoid the conclusion that six billion years would not suffice for the creation of man. However it is virtually certain that the Earth is not so old. It is a serious argument in favor of finality: the marvelous world in which we live cannot be the consequence of chance.

I add that all I know about the advances in biology confirms this idea to me. I previously talked about the progress achieved by biology when the role played by the ovule and sperm in reproduction was discovered. But where does this marvelous property come from to create a new being by the union of these? How are they made in the bodies of men and women? Who organized this astonishing path of the sperm encountering the ovum? Even if some day we answer these three questions, the reply will lead to others, and we shall end up by saying: "we cannot imagine that chance could succeed in doing all that."

The phenomenon of the healing of wounds leads to an analogous conclusion. Nowadays we understand the process well. But how does it happen that the elements necessary for the healing process are prescribed, that they hasten from all parts of the body, and know what they are to do? Is there not a marvelous organization?

I could examine the different functions of the human body. How does it happen that we sense hunger and thirst when our body needs to eat and to drink, that taste assures a first selection of nutrients, that our digestive system, after having treated these chemically, chooses molecule by molecule that which should be absorbed by the blood, that the heart, indefatigable motor, sends the blood everywhere that it is needed, and finally the lungs and the kidneys constantly clean the blood? It would be easy to continue this enumeration. Everything is a miracle. And the greatest miracle is the brain, in which chemical reactions become sensations and thoughts.

In thinking about these things, I was persuaded little by little, that the ideas at which I had arrived when I was between the ages of twenty and thirty, needed to be reexamined. It was necessary to acknowledge a certain final goal; at least it seemed to me that scientific progress made its existence very probable. But we are then at the heart of an incomprehensible mystery. There is nothing that we can affirm.

We can believe that the idea of finality would have led me to the idea of God. We could not believe in a final design other than by a conscious effort toward a deterministic goal, and if there is a being conscious of this effort, is it not God? But perhaps this impossibility of otherwise conceiving the goal comes from the weakness of our intelligence. Looking at a cataract, it provoked me to speculate: "Why do all these drops of water feel compelled to run into the sea?" I know however, that there is no conscious volition whatsoever. Perhaps in the same way, we are tempted to explain the goal by a conscious wish, even though there is none. No biologist will affirm that these astonishing molecules which possess the blueprint of life, are impelled by a conscious wish.

I often asked myself if the necessity of the laws of nature is not comparable to the necessity of the laws of mathematics. We do not refer to God in order to explain that the sum and product of two integers are independent of their order. It could not be otherwise. (On this point I am in agreement with Kant). Perhaps it is even impossible that the laws of physics, chemistry, biology be other than what they are. We cannot understand the reason for this impossibility. But can we deny it?

Thus I repeat, we are and remain in complete mystery. But is this a reason to give accounts which explain nothing? A name given to a mystery is not an explanation. Moreover, whatever good has been achieved by religion, the crimes committed in the name of God prove, in a singularly convincing manner, that none of these speak in His name. My conclusion then is that we must resign ourselves to the fact that we cannot elucidate the mystery of the world—a fact that was expressed by Pascal in one of the beautiful passages in the *Pensées*.

I am compelled now to remark that there is no arrogance in this conclusion. Believers speak of the arrogance of atheists. I believe on the contrary that it is they, who claim to speak in the name of God, whom we can accuse of being arrogant. In any case, to those who continue to think that God is the only explanation possible to explain the mystery, I advise them not to seek to define Him in too precise a manner. Each individual would have an idea, and we should perpetuate the era of the wars of religion.

To end, I would like to give my complete thoughts on the subject of religions. They have historically been instruments of good as well as evil. On the one hand, I see religious persecutions and religious wars; history is full of them. How can we forget it? I see Pascal too and all those who lived in terror because of fear of hell. But we should not forget all those for whom faith contributed to making them happy or at least less unhappy, nor all those who were led to help the needy. One should also think of all that which the arts (painting, sculpture, architecture, music) and poetry owe to religion.

On the other hand, in using terror to bring his readers to Catholicism, Pascal seems to me to be far from a true Christian. I often think that God, if He existed, would call Pascal's soul to court in order to say: I wanted to draw men by love. I am a benevolent and merciful God and did not contemplate the creation of hell. Those who imagined that have slandered me. I disavow them and I disavow you who have believed and sought to bring me men by terror. But even to punish them, I shall not create the hell that you imagined. You shall return to the state of denial to which you were, in any case, destined.

How do we weigh all this on the same scale? On the other hand one should observe that if there had not been religious wars, men would have other reasons to fight one another; the regimes of terror, to which, even in our day, many countries are subjected, do not seek justification in the will of God. In the same way, if religious art had not existed, secular art might have taken a direction that we cannot imagine. One cannot say with certainty whether religions have done more good or more evil. Each individual will judge it according to his own temperament.

As for me, I tend to believe that taking everything into account, in the last thirty centuries religions have done more harm than good. But they evolve; Christianity above all, has had a fortunate evolution. It has moved far away from the popes who preached crusades and approved the inquisition, or from an Alexander Borgia, to saintly persons such as pope John XXIII and Paul VI.

The church however seeks to turn attention away from this evolution. How can we testify that we have evolved, when we claim to speak in the name of God? Moreover do we forget that the dogma of infallibility of the pope dates from the nineteenth century, and that of the assumption of the Virgin Mary is even more recent. Moreover, for several years Catholics are taught to admire the continuity of the Church which has passed down the flame of the faith as well as Christian charity, from generation to generation. This continuity has not existed; we search in vain for the flame of the faith in a Borgia and that of Christian charity during the entire period from the beginning of the crusades until the end of the religious wars. On the contrary, the evolution of which I have spoken exists. I believe however, that it is the effect and not the cause, of the moral evolution of humanity, which has been brought about by French philosophers of the eighteenth century and by many others. Without doubt this evolution included set backs; our generation has seen more crimes than those that have preceded it. But in other times, criminals were cynical; in our days, they hide, or seek pretexts to justify themselves, and deceived by these pretexts, men who are entirely above suspicion, come to justify their crimes. But no one openly disputes the principles of morality: "Hypocrisy is a tribute that vice renders to virtue."* Perhaps virtue makes progress. I shall try to be optimistic.

Taking into account this evolution which is undeniable, and despite the Judeo-Islamic conflict (which is both racial and religious), in the coming decades, perhaps even longer, religions can do more good than harm. There is no benefit in accelerating their decline, which however, I feel is very probable.

To summarize the preceding, I believe, in thinking of religions and science, that "science will kill religion." Evidently this is not apparent at the present. One might have believed, after the world wars, people would have said to themselves: "God, if he existed, would not have permitted these to happen." I believed, and it should be testified to, that it is the contrary that took place. In their malevolence, people needed consolation, and persuaded themselves that religion would give them sustenance. I believe nevertheless, that the future will be without religion. Without truly proving that God does not exist, science will end by developing the critical spirit in men and to get into the habit of doubting all that is not absolutely certain. Perhaps there is interest in that which people of good will accept at the present time, and understand the necessity of struggling for the victory of a laic morality which is destined to replace religious morality.

* As to Pascal whom I consider one of the greatest geniuses of his time I would like to express some thoughts on his famous wager which has always inspired me. In the first place the problem is badly posed. I do not have only the choice of atheism and Catholicism: there are many other religions. Who can say if I converted to Catholicism after having read Pascal, that I would not fall into the hands of an irritated God who expected of me a pilgrimage to Mecca or the Ganges River? Ed.

The simplest schoolboy is now familiar with facts for which Archimedes would have sacrificed his life.
— Ernest Renan (1823–1892)

Goethe and Mathematics

Wilhelm Maak

Wilhelm Maak (1912–1992) was born in Hamburg, the only child of the banker Wilhelm and his wife Erna Solje. His parents were broad-minded intellectuals and when Wilhelm expressed an interest and inclination towards mathematics, they enthusiastically endorsed his wishes.

His early education was in the local schools in Hamburg and, at the University of Hamburg, he came under the influence of the noted mathematician Erich Hecke. When he was 20 years old however, he spent a year at the University of Copenhagen where he was strongly stimulated by Harald Bohr, the creator of almost-periodic functions and brother of the physicist Niels Bohr. In this discipline, Maak found the principal theme of his life's mathematical work. His contributions to this field, as well as others, quickly earned him an international reputation. He was an expositor of exceptional skill both in written and oral form. His books on almost-periodic functions and on calculus have been widely praised. He was said to have been a spellbinding lecturer both at an elementary as well as advanced level.

The period after the onset of the German catastrophe was a difficult one for mathematicians in Hamburg and no less so for Maak.

A very versatile mathematician, he supervised a large number of doctoral students and in the words of one of his students, Maak led them not only along the path of science but along the path of life as well.

He had auxiliary interests in the history of mathematics as well as very wide cultural interests. In 1939, he married Trude Lepper who shared many of his interests and their home became a hospitable gathering place for stimulating discussions. Visitors to their home included philologists, philosophers, and theologians.

In unpublished manuscripts, Maak wrote on mysticism and on St. Augustine. He also lectured on "Mathematics and Reality." The article on Goethe incorporates some of these observations.

He went to Göttingen in 1958 where he remained the rest of his life. He retired in 1978 and died in 1992.

Editor's Preface

In this essay Prof. Maak reviews Goethe's attitude and opinions towards mathematics and adds his own beliefs and convictions. Maak's principal thesis is that mathematics is a *language* and he claims that this is also Goethe's view. This language is used to communicate ideas that are not accessible in the normal vocabulary of everyday usage.

To study the meaning of words is a subtle undertaking and a great deal has been written on the subject. Even probing the meaning of common words is fraught with ambiguities and pitfalls. The meaning, for example, of such a word as "cognition" is not immediate, but evolves with time and usage. Its significance is amplified and clarified as we penetrate more deeply into the source and usage of the word. Indeed it often takes on a personal realization. To choose a widely used example in mathematics, the word "group" encapsulates a whole world of ideas whose complete significance is never fully comprehended.

Before considering the content of the essay, we say a few words about Johann Wolfgang von Goethe (1749–1832). His dates indicate that he straddled the age of enlightenment. While he is best known for his plays and poetry, he considered his scientific contributions at least on a par with his literary ones. With the passage of time, his science has been largely discredited but he clung tenaciously to the idea that to each phenomenon, there is an archetypal one, that is, an original type or form ("urphenomenon" in Goethe's terminology) from which later phenomena evolved. We encounter this same idea in Kant who wrote about "noumena" as contrasted with phenomena. The germ of both of these philosophical concepts is to be found in Plato. For example, behind a physical phenomenon, such as gravity, lies a mathematical formulation which may be a partial differential equation. But from these "primal" phenomena other physical phenomena may arise. Maxwell's equations are a good example—they are a "primal" phenomenon that govern several physical phenomena. This, however, according to Goethe, does not complete the story. The mathematician, for example, needs to go beyond that to find the source. This process alas, does not terminate, for the mathematician is compelled to descend and ascend again and, as the mathematician Hecke has remarked, this process never ends, that is, we never reach a complete understanding.

Maak quotes Goethe on the distinction between the objectives of mathematics and physics. Physics he says must pursue its goals independently of mathematics. These goals are the penetration into the sacred life of nature. Mathematics however, must pursue its objectives independently of all actual external influences. Maak notes that Goethe has understood well the structure and goals of mathematics. It does not receive its stimulus from the external world. Goethe is willing to acknowledge its independent course freed from all worldly contacts and it is then free to create its own world of concepts, structures and so forth. We add parenthetically that von Neumann, and others, are decidedly not in agreement with this point of view.

On the contrary! The art of the great mathematicians lies in phrasing the mathematical concepts and presentations in such a way that it permits them to make these their own, so that they conquer them and feel satisfied in their representation.

The vision that the mathematician forms for mathematical concepts cannot be different from those that pertain to himself. Released from outside influences, mathematics is then in a position to seek the primal source. As we noted above, it is never found. In the process of the search, however, a deeper understanding is achieved.

In any event, ultimately, says Maak, what activity one pursues is an individual matter but the goal should always be directed towards the interests and benefits of humankind and no discipline, he says, is more suited to this goal than mathematics.

The essay ends with a quotation form Goethe (which curiously, is almost identical to one from the English writer Alexander Pope)... "the proper study of mankind is man."

Every individual mathematician will periodically pose the question: What, in reality, is mathematics? The answer to this question presents an unusually difficult problem. I do not consider myself capable of satisfactorily answering the question. In fact, the published views and opinions concerning mathematics and its essence, which are disseminated, are mostly of an indecisive type. Even the great poet Goethe was inclined toward a prejudicial view of mathematics and mathematicians. In many respects he had, in reality, recognized the essence of mathematics without being aware of it and I thought that it would be interesting to hear what a contemporary mathematician had to say about Goethe's views. I hope that I am allowed the opportunity to give an interpretation and a critique of the statements of Goethe who gives us a transient insight into the world of mathematics.

"As one never questions the advantage that the French language has as a court and world language, which acts as a cultural and educational medium, so it never crosses one's mind to show disrespect for the virtue of the mathematicians in which they, in their language, cope with the most important knowledge acquired in the world, since they know how to identify and describe all topics that are subject to number and dimension." [In this and what follows, the passages in quotation marks are Maak's citations from Goethe's works. Ed.]

I do not believe that this quotation that asserts that mathematics is a language, was widely held in Goethe's time. Even today, many mathematicians retreat into silent musings if you say to them: Mathematics is indeed really a language. I claim that one can view other phases of mathematics than those we hold today and then perhaps the nature of mathematics can be characterized in another way. But the striking characteristic appears to suggest that it is indeed a language.

One of the greatest wonders which the mathematicians experience in their present status, is the almost ceaseless dual understanding, which mathematics in its essence makes possible. Naturally this indecisive understanding is purchased with many reservations. There are many which are not suitable for mathematical verbalization. Those parts have been relegated to the domain of mathematics. What remains can all be completely communicated. The restriction to the communicable part comprises its abstract element. But nothing is more misleading than to believe that mathematics is dull, that it consists of nothing but calculations.

On the contrary! The art of the great mathematicians lies in phrasing the mathematical concepts and presentations in such a way that it permits them to make these their own, so that they subdue them and feel satisfied in their reformulation. The vision which the mathematician forms for mathematical concepts cannot be different from those which pertain to humanity in general. These are formed through the history of the relevant concepts —e.g. through the pattern and concepts acquired through abstraction from perceived appearances of our world. In dealing with certain aspects of his science, such as tacitly representing them, how the mathematician's interests are formulated, etc., these matters are almost impossible to express in words of the German language.

The unique language with which to transmit correctly, the contents of his thoughts appears to be the language of mathematics itself. There he is certain that he will not be misunderstood, assuming that all participants are familiar with the language.

One sees the difficulties that are encountered by a young person when he tries to master the first semester of his mathematical lectures. A language is used there which he does not know. Once again he is in the position that every individual, as a child, goes through when he learns his mother tongue. Only through familiarity, and much determination, as well as imagination is it possible to master mathematics in all its intricacies.

I should like to recount an experience. In a lecture in the theory of numbers, I learned the concept of an ideal. In mathematics, an ideal is a set (usually infinite) that contains the sum $a+b$ whenever it contains a and b and that contains all multiples of c whenever it contains c. In the set of all integers, all even integers form an example of such an ideal. What image does one have whenever one speaks of an ideal? Certainly many think of an ideal as a bag full of number symbols and I can certify that this representation is not useful. After long speculation as a student, I came to the conclusion that in the case of certain ideals, viz. the prime ideals, they can be represented as none other than a small beer mug. I was delighted with this realization, and felt so certain that I went one day to the professor and said to him: " Should one not in a lecture, besides the pure facts, mathematical definitions, theorems and proofs, also note how the concepts have been secretly represented? What image should one form of a prime ideal"? He answered: " It is actually unthinkable for me to say in public, that I visualize a prime ideal as an unchewable ball and an ideal as a chewable one. If I were to say this in public, I would be considered crazy!" One should, without argument, admit that there can be no greater difference between a mug and a chewable sphere but this difference disappears if one restricts oneself to mathematical idiom. This conversation left an unforgettable impression upon me. It gave me the courage to fantasize within mathematics.

Let us now suppose that a non-mathematician speaks to a mathematician about a topic in a language he is used to. Certainly one would often have the occasion to encounter Goethe's experience:

"Mathematicians are like Frenchmen: whatever you say to them, they translate it into their own language and thereby it is immediately something different."

In a conversation with the Chancellor Muller, Goethe once said: "Mathematics is entirely false in the claim that it provides infallible conclusions. Its entire certainty consists of nothing but identities. Two times two is not four, it is still two times two and, for short, we call it four. Four however is nothing new. Thus does it persist in its development, except that in the advanced formulas, the identity is lost from sight."

"The Pythagoreans and the Platonists thought that all consists of number, even religion, but God must be sought elsewhere."

If one concedes that mathematics is a language, then the question arises, what subject matter can arise and be dealt with? Mathematics has the unique quality that it can bring into consideration concepts unique to itself. That holds for example, in Number Theory. One often hears the criticism: True mathematics, for example Number Theory, is indeed nothing but a tautology. Goethe said, "its certainty is nothing but an identity." Among other things one can concede that mathematics is a tautology. But one should not say that it is *nothing* but a tautology, one should correctly assert: mathematics is an insight into a tautology, and that is quite different from being *merely* a tautology. Goethe himself mentioned that in advanced formulas the identity disappears from one's sight. But that is indeed the responsibility of the mathematician, not to lose sight of the identity, in other words, he must master his topic completely. Only then can he achieve the insight to achieve complete understanding.

It appears as though mathematics has purchased dearly its certainty and its other properties, since it deals only with banalities such as the assertion that $2+2=4$. Such an assertion, even when it is correct, would not be designated as a mathematical piece of knowledge. We arrive at an extraordinarily interesting characterization of mathematics as a science. Before I proceed, I shall give two further examples which are of the type $2\times 2=4$, hence not to be viewed as mathematical concepts, although they are not as trivial as expressions such as the equation $2\times 2=4$.

One knows that the number $2^{257} - 1$ is not a prime number, that it can certainly be factored. But up to the present, no factors are known. This is the largest number of which such a statement can be made. [Since the appearance of Maak's article, this record has been greatly exceeded! Ed]. The determination of the factors, however, presents no problem. One needs only to sit down and attempt one integer after another to see if it is a factor of the given number. In a finite number of trials, we must certainly hit upon the factor sought after.

Another fact is the following: $2^{127} - 1$ *is* a prime number. This number has 40 digits and is the largest known prime [as of the date of Maak's article, Ed.].

However interesting these facts are, the mathematician views them as curiosities but not as mathematical insights. All these facts can be determined in a finite number of steps; thus with sufficient endurance, anyone having sufficient time can verify the statements and obtain the factors. Whatever this is, however, it is not mathematics.

By contrast, the famous theorem of Euclid, "There are infinitely many prime numbers," is an authentic mathematical theorem. By trial and error, one can never verify this theorem. In contrast, if a person, going through the number series finds billions of consecutive numbers which can be factored, we cannot then come to the conclusion that there are no more primes. That would contradict the theorem of Euclid. Its proof requires a method other than that given above which investigates a single integer. And this methodology can be viewed as characteristic of real mathematics.

Many mathematicians are in agreement that mathematics is in reality the science of the infinite. Thus to genuine mathematics belong only those statements in which somewhere, the infinite plays a role. The expression $2 \times 2 = 4$ is thus not a mathematical statement, and the same is true of the statements $2^{127} - 1$ is a prime and $2^{257} - 1$ is not a prime.

Frequently one hears the point of view: "you are a mathematician, so you must be a good chess player." In fact chess has essentially nothing to do with mathematics. The construction of a theory of chess, which views chess entirely as a game, is a problem to be viewed in the same way as whether $2^{127} - 1$ has a divisor. It can thus be determined in a finite number of steps. For card games such a theory, which in principal can be given for every board game, is impossible. If the card games have more the character of a game than chess, the mathematician comes to the same result as, for instance, E. A. Poe in his interesting investigation which one can read in the introduction to his thrilling story.

It is thus the "infinite" whose consideration fascinates the mathematician. A contemporary mathematician would never seek an individual number in which one finds mystical interpretations. But in earlier times, it was otherwise; numbers were identified which served as symbols of essential content. Thus were the so-called perfect numbers, e.g., the number 6, as well as the pairs of amicable numbers which were symbols of complete and friendly persons. In the symbol 1, one found the parable of God. In fact, according to Goethe: "God must be sought in another place." However to grasp the infinite, my intuition suggests an incursion into mystical regions. And we are wonderstruck when great mathematicians succeed in making assertions about the infinite which are inconceivable to us.

What can we say about the claim about the certainty of assertions that mathematicians make about the infinite? We saw that a theorem such as Euclid's quoted above, is in no way as trivial as that of $2 \times 2 = 4$. The theorem is no longer viewed as being a mere identity. And in fact, doubt about the complete trustworthiness of the conclusion is not unconfirmed. In general, where mathematics is essentially interesting, its conclusions are forgotten and hence, among other things, doubtful. This powerful fact was achieved at the end of the last century as a mathematical conclusion, since in fact one thought about the infinite carelessly. One of the great achievements of Hilbert was the recognition that it is necessary, in achieving a proof, that any assertion about the infinite must be made with great care in order to achieve reliability. In general what does it mean to say: "To make reliable claims about the infinite," since we are indeed precluded from ever experiencing the infinite. Even this question was satisfactorily answered by Hilbert and in fact a part of mathematics, for example number theory, is available for the verification of Hilbert's assertion.

One of Goethe's aphorisms can be used to summarize our views on this point: "One hears that mathematics is certain; it is no more certain than other knowledge and action, it is certain if it knowingly deals with things about which one can be certain.... In this sense, one can speak of mathematics as the highest and most certain knowledge."

We suggest that one of the most pressing tasks of the new mathematics is to limit the domain in which we can work with absolute certainty. Goethe said: "A benefactor of mankind is one who can teach a critique of human understanding; to encompass human understanding within his sphere of knowledge."

Henceforth, we consider those figures which in addition to numbers, inspire human thought. Up to Goethe's time man lived in the supposition that the study of figures, that is geometry, was a type of natural science, namely the science of the motion of things surrounding our world. This assertion was found to be false at the beginning of this century. In geometry we can immediately pursue a development that supports our thesis that mathematics is a language (in contrast with mathematics as a natural science).

In a conversation that Goethe had with Chancellor Müller, I find the following report, "From 5:00 till 8:00 o'clock, I visited with Goethe, whose conversation was most interesting, thoughtful, and warm. He spoke of Cuvier's eulogy on Hauy, where we find 'the heavens are entirely subject to geometry.' Goethe smiled mockingly at this phrase since mathematicians could not even clarify the centripetal force."

The conversation took place in 1823. Precisely in this year, Bolyai showed for the first time that the parallel axiom cannot be proved. He showed that indeed a geometry can be conceived, in which all the Euclidean axioms can be verified except for the parallel axiom.

In his famous inaugural lecture Riemann showed, among other things, that indeed in our world a noneuclidean geometry can be realized. He proposed, among other things, experiments which could determine the question as to which properties our universe possesses. The working out of Riemann's idea was carried out recently with Einstein's relativity theory whose most essential thrust was a renunciation of the claim that our space is Euclidean.

The greatest significance of this development is in our realization that geometry is not concerned with the characteristics of the surrounding space but with thoughts and suppositions which, to be sure, can be used to describe a more or less applicable description of real relations. In a remarkable way for example, relativity theory makes possible, in its language of tensor calculus, an understanding of centripetal force as well as gravity even if it does not provide a complete clarification.

Geometry considered as a science, is a science of the possible, not properly of reality. It is in a position to demonstrate possibilities that can arise in nature. To decide which of the possibilities are verified does not lie in its realm of responsibility.

Mathematics, and especially geometry carries a great educational significance. So claimed Moltke. As examples of its utility, mathematics is an instrument and in a position not only to calculate the trajectory of projectiles, but by the intellectual training which it uses, especially in its properties, it is the science of the possible. Even Plato had the conviction that education in mathematics must be the basis of the rearing of a statesman.

I believe that the educational value which Goethe ascribed to geometry, arises more in another sense than Goethe assumed. In Wilhelm Meister he wrote: "When a child begins to understand that a given point does not have a predetermined predecessor, that the next path between two points should be thought of as a line, even before he traces it with a pencil, then he experiences a certain pride, a relief. And not without justification. For then is the source of all thought opened to him, ideas and realization 'power and action' become clear to him. Philosophy discovers nothing new for him, the geometer on his part initiated the basis of all thought."

According to our perception, the stage which has been reached does not appear to be the realization of an idea. Moreover, it concerns a more or less appropriate pictorial correspondence of the objects thought about and the objects of the real world. And what Goethe's child can learn is the existence of all human knowledge: that includes the world of phenomena concerning our most fundamental existence, in which case our most basic world of mathematics.

The most familiar non-mathematical discipline similar to mathematics is mathematical physics. Goethe's statements on this branch of mathematical activity are particularly well known; since with shocking stridency he has spoken out against the application of mathematics in many branches of physics, for example he wished to prohibit its use in the theory of color. Helmholtz had once referred to a particularly pithy expression of Goethe on the debatable part of color theory, which refers to Newton's mathematically oriented color theory. "Unbelievably brazen." "Highly appropriate for the students of the left bank," "grotesque clarification," "perfect nonsense," "But I see clearly, necessary lies to the masses." You all know how much Goethe erred in his proclamations; since despite Goethe's incantations, mathematical physics made further progress and celebrated triumphs of a specific type in radio, atom bombs, and other achievements of recent times, which would not have been realized without the use of mathematics.

I found Goethe's assertions repeated in the following lines in more tempered and objective tones: "The following must be sharply distinguished from one another: physics from mathematics. The first of these must function in a decidedly independent way, and with all enchanting, respectful, reverential, powers in nature and the changing life, to seek, entirely unencumbered, what mathematics on its part, directs and does. The latter must however, on the other hand, declare its independence of all external appearances, it has to follow its own magnificent spiritual path and more purely develop itself than is possible, so long as it tries to deal with reality, and with the means available, cedes and takes some pleasure in it or endeavors to assimilate it as it really is."

A brief suggestion from Goethe's setting on natural science cannot be omitted at this point. One should indeed claim that he represented one of the last and greatest scientists of a past epoch. Goethe appears as luminous as the botanist and systematizer Linnaeus. Goethe's natural science (as opposed to mathematical physics) can perhaps be characterized as a reconsideration of the structure of natural phenomena with concepts of ordinary speech. That applies for example even far removed from Goethe's theory of color. One can understand it as a most natural systematic structure of color, similar to what modern botany gives about plants. Theories in our present day sense are almost completely forbidden in Goethe's science, and where they appear are unassailable. This renunciation of theories lies at the basis of a very deep understanding of Goethe. The spoken and written word has a very responsive life. It dies or it lives in its own way if it does not remain in lasting, continual contact and inner relation with our surrounding reality. In general where the words distance themselves from the living reality, for example in rambling theories, their significance becomes uncertain, or indeed it becomes mēaningless to function with them. Therefore Goethe was of the opinion that self-evident theories are repugnant; the theories must be derived from the phenomena themselves.

Besides he sought to justify the individual significance of words. So I found in some place, a praise of those words which like the word "bang," give the concept from their linguistic representation. Likewise he is very pleased with the fact that words having a similar significance such as "mein," "dein," "sein," sound alike while he finds it unfortunate that the words "Ich," "Du" are so dissimilar. It is better in French where these are "Moi," "Toi." He finds it particularly fortunate in finding words that have a resemblance to their branch of science such as "polarity" in the theory of magnetism, and "affinity" in chemistry. Here the significance cannot so easily be changed as is the case with other words.

Goethe wishes to undertake speculations on nature to see to what extent primal phenomena ("*urphenomena*" in Goethe's terminology) lie at the basis of our understanding of natural phenomena. The significance of the word (*urphenomena*) is clear when we compare it with similar sounding words. Very close is the association "primal mother" ("*urmutter*"). In fact the following helps us to an understanding: if one sees in the grandchild characteristics similar to those in the *urmutter*, then we infer that they have a common ancestor. The primal plants, primal animals of Goethe are thought of in the same way. We are dealing with an understanding of the properties of a primal phenomenon which in experiences of the world are certain, as for example, the common ancestor to the descendant. Goethe considers in this way all of nature even, for example, color. Goethe does not admit that a primal phenomenon is self-evident, independently of its appearance, and yet meaningful for the researcher of nature. We must, in Goethe's opinion, be content with it, to have confidence in it, to experience it more and more in the manifestations of the world.

Another association: *Ursache* (causes) divert us from the appropriate way. Or even more, the word "*Ursache*" is to be understood in apposition to 'cause.' Researchers in Goethe's time and even after, sought for the causes of phenomena. That is what Goethe struggled against. In Goethe's opinion a knowledge of the world requires not uncertain causes but requires us to see in the evidence primal phenomena, which are certain.

In fact it is likely that the physicists of earlier times moved along false paths while they sought the causes of phenomena, from which they later derived the phenomenon itself as a consequence. At least our present attitude has altered considerably. As physicists use mathematics more, they nevertheless have a clearer understanding of the point of view that agrees with Goethe's perception. Today they are all, in addition to Goethe, of the opinion that one should not clarify natural phenomena with theories but should describe the phenomena. Faraday and Kirchhoff were the first who advocated this point of view particularly clearly: one can arrive at a knowledge of nature only through observation. The result of these observations must be recorded somewhere. Goethe uses the German language. Modern physicists prefer mathematics.

In which way however, is mathematical language, in comparison with ordinary language, so well suited to comprehend an image of nature? The answer is: mathematics has an extraordinarily developed structure. It is also unrelated to the real outside world. Its concepts are, moreover, unlike those of ordinary language. While ordinary words have meaning only in relation to the outer world, mathematical concepts have an essential existence in themselves. It is well substantiated that mathematics, as we exhibited in detail for number theory and geometry, can itself choose what objects it treats. Therefore its concepts and symbols have had a well-grounded and unvarying existence, independent of the external world, for thousands of years. And this extraordinary property of mathematics has been used by natural science with outstanding success. A typical example will clarify this. If one describes the phenomena of the world, then one can, in case our esthetic sense demands it, add good, but not directly observable characteristics, without running the risk of being lost in vague speculations.

For example the Maxwell equations are a sort of extension of experience. They prove their value in an entirely unexpected way in that they make it possible to describe in a unique way, not only the theory of electricity, but also the theory of magnetism as well as optics. Helmholtz once claimed that the Maxwell equations must be rightly described as a primal phenomenon. Whether Goethe would have been in agreement with this point of view is a question that I shall leave aside. Perhaps it would have had the consequence that the success of the mathematical description of nature would have changed Goethe's mind. I surmise that this would have been very unlikely. He probably would have been more cautious in his criticism: except for the fact that the Maxwell equations are not, in Goethe's sense, certain. In that case, then his attitude would be incapable of being justified. He was completely incapable of following a long mathematical argument. In a letter he once wrote: "I am dependent on words, language and image in their proper meaning and completely incapable in any way of operating with symbols and numbers that a highly talented individual easily understands."

I am of the opinion that in fact, the tension relation "phenomenon-primal phenomenon" is the driving power of every scientific activity. Wherever there appears a phenomenon in the world, it is nothing more than a manifestation of a primal phenomenon. If that were true, and if in fact mathematics, correctly applied, is in a position to give, in its details, a suitable image of the physical world, then we should be led to the conjecture that the relation phenomenon-primal phenomenon must be mirrored in mathematics. And indeed this is a fact. One soon learns that the deepest insight can only be achieved if one does not restrict oneself to a consideration of the objects of the domain, to work with them, to calculate and to operate. One must rather be in a position to view these objects as unique manifestations of a super-ordered primal object.

An expert in calculations is not a mathematician. He is a mathematician if he is in a position to look beyond what the calculation has accomplished. Many different functions which are connected to a phenomenon, satisfy a differential equation which represents the primal phenomenon to which it belongs. Progress in mathematical knowledge is obtained if we forget the phenomenon and bring into consideration only the primal phenomenon. These will thereby appear, in their own right, to the naked eye as phenomena and the mathematician will feel himself compelled to ascend or descend to the higher or lower regions of the primal phenomenon. In fact noted mathematicians such as my teacher Hecke, have pointed out this step-like progression of the mathematical world. The unlimited progress in the higher or lower regions conceals the hazards. With every step one believes that he is coming closer to the absolute truth: Alas this is not the case. The abandonment of phenomena in favor of the primal phenomena of primal phenomena leads to ever clearer but less vivid regions of mathematical existence. One can never bring mathematics to an end; around 1927 it was proved that in every well-defined mathematical discipline that includes segments of the mathematical world, one can always find truths which cannot be verified within this discipline. In order to see through it and to prove it, one must rise to another sphere of existence. So mathematics recognizes a level higher than itself.

It is necessary once again to point out Goethe's attitude toward primal phenomena. It differs in a vital way from a typical mathematical tendency. Goethe, as a researcher into nature, forbids the assignation of primal phenomena as being other than phenomena. An ascent to higher regions of the primal phenomena appeared to Goethe to be arrogance. Goethe's opposition to mathematicians is thus not only a consequence of his limited ability in mathematics, but his opposition is based on a world view (*Weltanschauung*). It cannot be denied that in unlimited "mathematizing," there is a danger of losing one's self. This is a fact worth noting.

Experience shows that mathematics is well suited to give a general truth about the world, i.e., a consistent image of the world. I personally have the view that even the mathematics of integers, on the basis of its restricted structure, represents an image of the world: a symbol whose existence in us can generate a feeling that we can better understand the world, or its existence.

We have seen that Goethe had included mathematics in his thinking even if superficially. Natural science in its entirety had occupied him for decades. One hears the complaint that he should have written poetry instead. As an alternative to the many volumes of natural science he should have written more plays and songs. One can only be led very superficially to such a wish. Goethe's works, such as *Faust*, would never have had the deep content, if Goethe had not attained an insight into the different realms of human life. Above all in the second part of *Faust* one finds in different places, and precisely in these, the deepest arousal of thought and feeling as well as woven perceptions, which Goethe had arrived at through his work in natural science. I recall only Faust's relationship to the mothers.

Correspondingly, in a certain sense but conversely, it is the same with a mathematician. He is also in a position to carry out the deepest things mathematically but yet not forget that he is not only a mathematician. Moreover, he must know in what way to begin mathematics, this unbelievably fine structure. Mathematics is, in the final analysis, also a language. It is not useful if one can already speak, and it is not to say "The mathematician is only as complete as he is complete as a man. When he senses in himself the beauty of truth, then he will be thorough, lucent, circumspect, pure, clear, charming, and act elegantly. All these things are needed in order to be similar to Lagrange."

Lagrange is a mathematician already far removed from us in time. The most brilliant example of a mathematician is for me Riemann whom I have already mentioned as the pioneer of relativity theory. He is probably the greatest mathematician in general. And Riemann was not only a mathematician as Goethe was not only a poet. Riemann placed great value in his philosophical speculations. It was for me a great experience when, as a supplement to his work, I learned of these. They gave me the tools to understand Riemann's language in a deeper sense, as the language of an ideal person. He was deeply religious; he did not envision his mathematical powers as a personal gift, rather he thought of it as a part of the infinite intellectual content of the world which materialized in him. He found this stream so vivid that the idea came to him that the idea of the entire flow of intellectualism to the world could be adapted to clarify the force of gravity. I believe that these and similar marvelous ideas, and Riemann's towering prowess as a mathematician, are essentially related. The truths in his speculations are given to us in mathematical language free of any inconsistencies.

It is impossible to say what one should or should not do. It depends upon the fact that in the final analysis our activity should be an expression of humankind. So it is with mathematical activity which many people sense is a non-human endeavor. On the contrary, I should personally like to claim that indeed, there is no more human endeavor, no science more valuable to mankind, than is mathematics.

In Goethe's book "Selected Affinities," I find the following words: "To the individual remains the freedom to choose what attracts him, what gives him pleasure, what appears useful to him: but the proper study of mankind is man."

Nature and nature's law lay hid in night
God said "let Newton be" and all was light.
—Alexander Pope (1688–1744)

Leonardo and Mathematics

Francesco Severi

Francesco Severi (1879–1961), one of Italy's most renowned mathematicians, was born in Arezzo. The Tuscan country is rich in the traditions of the arts and sciences. His antecedents left the land and became active in several professions and in particular, his father was an active official in local affairs. His mother Licinia Cambi was reported to be a person filled with religious fervor and endowed with wisdom, courage, and habits of industry and thrift. There were nine children of whom only five survived. Tragedy struck when his father took his own life in 1889. Despite this and resulting financial hardships, the mother kept the family together and saw to the education of the children. Francesco helped the family finances by tutoring other children in school, even those older than himself.

He entered the University of Turin with the help of a scholarship, and in 1900, obtained a PhD degree under the direction of Corrado Segre with a thesis on the singularities of a curve on a hypersurface. He was then appointed to a position as assistant. Then in a better position to support a wife, in 1900 he wed Rosanna Orlandini, his fiancée of several years. Rosanna had the reputation of being a warm-hearted woman of great intellect. She was his companion for 52 years.

He held posts at Pisa, Bologna, Padua, and Ferrara. In Padua, in addition to his academic work, he held a civil service position in the gas and water works as well as director of the school of engineering. He was then called to the University of Rome where, in 1923, he was appointed rector. While in Rome, he had conflicts with the fascist regime. In 1939, he established the National Institute of Higher Mathematics.

Severi was a very prolific writer and was recognized for the originality and profundity of his work. The noted geometer Beniamino Segre, using artistic metaphor, described Severi's work as a grand fresco rich in ideas and content, which had a great impact on the development of mathematics in the 20th century. His work influenced workers in other countries as well as in Italy. He received many honors both at home and abroad.

He is said to have had an intense and complex personality that sometimes led to difficulties. After the unexpected death of his wife in 1952, and with failing health, he took a keen interest in philosophical matters as well as religious ones. He wrote on ontology and epistemology as well as on religion. In particular he was interested in matters of faith, which he hoped to solve logically and scientifically. He wrote, "It is absurd to deny that which we do not understand; it is absurd that one does not seek faith."

He was guided by one of his heroes, Leonardo, who wrote "As a well spent day deserves a good night's sleep, so a well spent life deserves a good death." So it was with Severi.

Editor's Preface

Prof. Severi's goal in this essay is to analyze and express an appreciation for Leonardo's excursions into mathematics.

Leonardo da Vinci was born in 1582 in the town of Vinci, close by Florence, the illegitimate son of a successful Florentine lawyer. He was reared in his father's villa.

He is generally recognized as one of the most, if not *the* most, original scholar-artist of all time. Although not strictly eponymous, the phrase "renaissance man" conjures up in many people an image of Leonardo himself or one aspiring to his cosmic intellectual and artistic achievements.

His most enduring work consists of his artistic accomplishments. They are few in number but have a stunning beauty, and although many of them are rooted in the Christian tradition, they transcend that association and have received universal appreciation.

His other work has come down to us through his notebooks. Leonardo was left-handed and sketched and painted with his left hand. His notebooks were written by Leonardo with his left hand and *backwards,* that is, in mirror image. Some have surmised that Leonardo intended this method as a code. But the argument is not convincing since the "decoding" consists simply of holding the text to a mirror. One author has suggested that Leonardo suffered from agraphia, a recognized writing disorder, and that this device was his way of circumventing his affliction. The argument, while plausible, is not persuasive. In any event his books and notebooks contain a wealth of material which, even to this day, has not been fully assessed. These materials are to be found in museums in many countries of Europe.

In the case of so prolific a writer, endowed as he was with an overwhelming thirst for knowledge, it is not surprising that many of his mathematical ideas were not carried to fruition. It is therefore important not to overstate his accomplishments and Severi is careful not to. Severi makes an honest assessment of Leonardo's mathematical achievements and, as he notes, some of these were profound, albeit incomplete. Severi points out that Leonardo was not a mathematician in as strict a meaning of the word as we should interpret it nowadays, perhaps because, as Severi notes, the mathematical environment in which Leonardo lived stressed the computational aspects of the subject. A creator of Leonardo's breadth did not have the time to elaborate his myriad ideas. In particular, he was especially anxious to bring his investigations in mechanics to fulfillment with the help of mathematics and this diverted him from his "pure" mathematics.

Severi gives a brief listing of Leonardo's mathematical accomplishments, to which a few are added, taken from Leonardo's notebooks—in particular the Codice Atlantico: (1) The construction of regular polygons. (2) The sum of arithmetic progressions. (3) The area of lunes. (4) Regular polyhedra, that is, the Platonic figures. He collaborated with Luca Pacioli on the book *Divina Proportione* which contains a large number of Leonardo's beautiful drawings of these regular solids. (5) The Delian problem, that is, the construction of a cube of volume twice that of a given one. This problem arose because of the desire to appease the gods by building an altar with twice the volume of the given one. The construction has long been known to be impossible within the limitation of doing so with straight edge and compasses alone. (6) Finding centers of gravity. (7) Mathematical analysis of the motion of fluids. (8) The use of "infinitesimals" to determine areas of figures. For example he found the area of an ellipse with semi-major axis twice that of the minor axis. He does this by cutting the ellipse into "hair-thin" parallel strips and comparing these areas with the area of corresponding strips of an inscribed circle. It is fair to say that, in this process, he had foreseen the basis of integral calculus. He did not have the advantage of seeing the "Method" of Archimedes since that work was not discovered until the beginning of the 20th century. (9) His disquiet concerning the nature of points, lines and surfaces. The line, it was asserted, is an infinity of points but Leonardo is uneasy adding a lot of "nothing" to get a line. In short he was very mistrustful about the nature of the infinite. This skepticism concerning the infinite reasserted itself five centuries later in the "intuitionists" of the 20th C. (10) He actually gave a sketch of a mechanical device to calculate powers of integers, in particular 20^{23}. Why he wanted this value is shrouded in mystery! (11) Given a spherical mirror and two points *A* and *B* outside, to find the point on the sphere such that a ray of light emanating from *A* is reflected from the sphere and strikes the point *B*. This is usually called Al Hazan's problem but Leonardo was unaware of the work of this noted Islamic mathematician.

An obvious question that remains is the extent to which Leonardo's mathematics influenced succeeding mathematical scholars. This influence is very likely minimal since many of his notebooks did not come to light until decades, and possibly centuries, later. On the other hand we see in his mathematical work another testimony to the profound genius of this extraordinary "renaissance" man.

Author's Summary. *Leonardo was not a mathematician in the strict sense of the word—certainly not because his genius could not also range over abstract fields, but because in the fifteenth century mathematics was cultivated primarily with a view to practical and concrete objectives; and it is to these ends that da Vinci's works and thought were primarily directed when he was not occupied with art. It was necessary for him to reach his goals quickly, as, for example, his applications of mathematics to mechanics. The fresh start on Greco-Alexandrine science had barely been inaugurated in the 15th C. and the famous school of the algebraists of Bologna was born when Leonardo was already old and peripatetic. Nevertheless, the great artist-scientist did not cease to reflect deeply on true and proper mathematical concepts (points, infinity, etc.). He brought to fruition several contributions to quadrature and cubiture, with a type of induction on infinitesimal methods; these elucidate the indivisibles of Cavalieri. He was, in any event, a fervent lover of mathematics, in which he found a sense of supreme certainty.*

Leonardo and mathematics, I wrote deliberately, rather than Leonardo the mathematician. A pragmatic spirit, (although, without wishing to be, and perhaps without realizing it) he was soaked with neo-platonism, a lover of natural reality, as with almost all the Tuscans of the Renaissance, and mathematics could not be, for him, anything but an instrument, both exalted and powerful, to understand the reasons for different phenomena. It is essential to know the causes, to be sure, but also to come to fruition as, for example, with mechanics.

"No effect in nature is without a cause…understand the reason and you do not need experiments…nature is full of infinite reasons which are no longer based on experience…. But "whoever claims, from observation, that which is not in him, has moved away from reason…observation does no good—only our opinions are of value."

He knew at least the work of Bacon [presumably Roger Bacon, 1220–1292, Ed.] through the book "Leonardo Vinci Disciple of Observation", his method was a precursor of the Galilean method; the mathematical apparatus was the secure bond between reason and experience. Intuitive reason predetermined the experiment, related the result, and subjected it to all criteria and inquiries, qualitative and quantitative. If this inquiry enables you to discover the necessary causes and succeeds in synthesizing them into law, proceed in another manner: you do not need observation any more. This procedure never fails because the rationality of nature is absolute.

In any event, Leonardo was not a mathematician in the strict technical sense of the word. It is not that lofty and deep thoughts did not please him. The powerful intuition of natural truth, which he almost palpably loved, carried him toward art and applications; his philosophical inclinations carried him instead toward abstraction and objective problems and from another point of view opened his interest anew towards art that was, for Leonardo, the place where all the activities of the spirit met.

He wrote that, the illiterate man, the disparager of liberal humanism, the inconclusive discussion, foolishly and admiringly copying the old, critics of the "trumpeters and reciters of the work of others"—all search avidly in the libraries for the oldest and most appealing works, especially in mathematics and philosophy, and note where this or that can be found: Euclid, Archimedes, Aristotle, and so on.

The problem of zero and infinity became for him a torment when mathematics was in the forefront of his interests. "The point is nothing and, "about nothing, it is not possible to communicate any knowledge."

You cannot create a body from something which has no substance." He therefore abandoned the static Euclidean conception of the line and surface, whose places are given by points; and as with pictures, he reacts to the stationary statues of his predecessors, even the great ones such as Piero della Francesca, and gives to the individual creations the features of movement and variety and variability, both of pose and attitude; so in geometry, he preferred the genesis of figures having motion; at least the body, which is concrete and pliant, and he conceives the surface as the boundary of the thing in question, the line as the boundary of the surface and the point as the boundary of the line. However, there is always the necessity to convert from the object to the abstract—pure creation of the mind. But the passage is difficult; "to go from the natural point to the mathematical point is an infinite jump."

Concerning the infinite he poses as well the question: "What is the thing which cannot be given and if given it could not be known." He replies : "It is the infinite." "The point does not have substance and is contained in continuous quantities" (formed of infinitely many points). "An instant does not have time…it resides in time (formed of infinitely many instants). Despite that, only the present exists in time, since the past is in memory and the future in imagination." All of which is reminiscent of St. Augustine.

But there are additional abstractions, which were induced by the philosophical nature of his intellect, perhaps against his will.

Leonardo's mathematical knowledge was not great but adequate to a time when the contact with classicism was filled with art; it was not yet permeated with science. Archimedes was not known to him except, strictly speaking, briefly. "The Archimedean quadrature of the circle" he wrote in the Atlantic Codex, "is well expressed, but badly realized. It is well expressed where he says that the circle is equal to a polygon made up of the segments of the circumference and the radii of the circle: and it is badly communicated when he gives the area of a circle that of a polygonal figure of 96 sides; this fails to take into account the 96 little pieces cut off from the 96 sectors." As was discovered only recently in our century, the great Syracusan, who was not a stranger to the fundamental concepts of infinitesimal calculus, knew very well that when Leonardo said "well expressed" to the theoretical expression for the exact area of the circle, and that the reference to an inscribed polygon of 96 sides was nothing more than a way of calculating the value of π approximately. But the loss of Archimedes' work was by now, several centuries old, and one cannot blame a person, who coming into contact with it for the first time, did not succeed immediately in interpreting its depth.

Also the work of Diophantus of Alexandria was momentarily forgotten and the same is true of most of the work of Leonardo Fibonacci of the twelfth century. The great school of the Bolognese algebraists arose at the beginning of the sixteenth century; but Leonardo, already traveling, could not have had close contact.

The principal mathematical studies of da Vinci, after those of his childhood with the Florentine abacists, were above all inspired by his relation to Luca Pacioli, pupil and contemporary of Piero della Francesca, one of the most famous mathematicians of the 15th century, professor at the University of Milan, Bologna, Pisa and Rome. From "master Luca" he acquired immediately, for 119 ducats, the book *Summa de Arithmetica* printed in Venice in 1494. In Milan too, in the period of the "Moor," Pacioli and Leonardo, in 1497, worked together on *The Divine Proportion*, a kind of obsession of the artists of the Renaissance, which they sought in the faces, in human bodies and finally in the lines of the swiftest animals. In the edition of this book which came out in 1509, the sketches are "made of the most worthy pictures, perspective, architectural, musical and all the virtues with which Leonardo da Vinci, Florentine, is endowed." Thus wrote the same master Luca.

Leonardo, according to his own intuitive qualities, liked geometry (with which he finally solved problems of arithmetic) more than arithmetic. This limited him as before to the basics (fractions and radicals, according to what he said), and these could not interest him very much. In any event he studied it and his calculations were uncertain, and at times erratic in his first works, but, in his later works, as certain and accurate as in a contemporary one. Of algebra (an *art* according to the contemporary terminology) he knew little.

Now we come to a quick enumeration of his geometric researches, with which he dealt, using elementary and primitive methods, in which he showed a profound ingenuity worthy of higher subjects than those he intended to study, perhaps because of the involvement of artistic or technical effects.

The quadrature of lunes, extensions of the lune of Hippocrates of Chio (a plane figure whose boundary is made of straight lines or circular arcs). Concerning these he found a theorem not known to him, but which had been discovered five centuries before by the Arab Al Hazan; transformation of solids into others, "without increasing or diminishing the volume" ("equivalent" solids as we should say today); also in relation to the classical Delian problem of the altar, the geometry of the tetrahedron; geometry of the compasses (anticipating by a half a century G.B. Bennedetti and by three centuries Lorenzo Mascheroni); the center of gravity of the point of geometric view; the problem of Al Hazan on the reflection of a radius in a spherical mirror (the problem of incidence as Leonardo expressed it). This question is not an easy one. He attempted it several times, without achieving his goal. It is a problem of the fourth degree, elegantly solved by Huygens almost two centuries later, but of which Al Hazan and Vitello had first given many most intriguing solutions. Leonardo could devise a solution, "using instruments" with an articulated parallelogram, which was constructed by Marcolongo in 1929, based on the availability of drawings of Leonardo.

We recall also his precepts perspective on the heel of Piero della Francesca, in the *Treatise on Painting*. For example the rule which he gave to painters for the distance of the subject in the background, shows the validity of the knowledge of the aperture of the optical cone of visibility for a fixed eye.

But the following are above all to be emphasized in the mathematical work of Leonardo, that is, the first hesitant, but significant, introduction of the method of infinitesimals, the glory of Archemedes and the 17th century. The attempts at finding the area of about 80 spherical surfaces by the division into thin strips and the quadrature of a particular ellipse as an affine (as we would say in contemporary language) transformation of a circle, a prelude indeed to the method of indivisibles of Galileo-Cavalieri, anticipating the infinitesimal calculus of Newton-Leibniz. "This is evidence," he wrote in the Atlantic Codex, "and remains convincing, imagining the circle to be divided into tight parallels in the manner of very thin hairs in continuous contact with them...".

Leonardo was an avid lover of mathematics in which "he found a sense of beauty worthy of music and of infallible things …Whoever denies the supreme certainty of mathematics will be repaid with confusion and will not be able to reduce to silence the contradictions of scientific sophistry, which make an unending noise." "Whoever does not question whether two times three equal more or less six…; on the contrary, the discussion will vanish into an eternal silence."

"No certainty exists to which one of the mathematical sciences cannot be applied, or at least which are not related to it mathematically." The other wonderful circumstance is that his mathematical research never proceeded independently from his innumerable other activities both technical and artistic. Some of these date from his journey to Florence in 1500, after he had left Milan, not only his research on the lune, but also his cartoons for the "Virgin and Ste. Anne" and for the "Virgin, Ste. Anne and Child," (Louvre); and from the journeys to Florence from 1503 till 1507, after the fall of the Borgias, the researches on the transformation of solids, the cartoons for the Battle of Anghiari, (for a variety of reasons this painting never came to fruition) and the celebrated portrait of Gioconda.

A man of the stature of Vinci could not have been unaware of the relative paucity of his mathematical studies. And in fact he wrote in the Atlantic codex: "Seeing that I cannot contend with material of great use or pleasure…I shall do as some poor person who arrives finally at the fair…and must be content with all the things others have seen, and rejected." An almost impersonal judgement worthy of his greatness.

Number theorists are like lotus-eaters—having once tasted of this food they can never give it up.
—Leopold Kronecker (1823–1891)

The Highest Good

Norbert Wiener

Norbert Wiener's (1894–1964) education and career were unconventional. This could probably be attributed to the fact that his father was a rarity. In fact, his father Leo, was born in Byelostock, White Russia and was a Mark Twain like character who, after becoming fluent in 40 languages, was appointed a professor of Slavic languages at Harvard by way of Missouri where Norbert was born. Leo married Bertha, née Kahn, a remarkable woman overshadowed first by Leo, then by Norbert.

Norbert was named after one of the characters in a play by the poet Robert Browning, of whom Bertha was very fond. (A daughter Constance, born later, was named after another character in the same play!)

Norbert was a child prodigy learning to read at age three and completely fluent at age six. His education was carefully directed by his father; Norbert graduated from Tufts and went on to complete a PhD in logic at Harvard at the age of 18. He then received a traveling fellowship that he used to study in England with Bertrand Russell and G.H. Hardy and then in Germany with David Hilbert.

Upon his return, he did not follow a standard career in academia but engaged in various occupations—journalist, engineer, editor, writer, and he eventually began a long and fruitful association with the Massachusetts Institute of Technology. MIT was emerging from its tradition as a high quality engineering school to a high caliber broad-based academic institution.

He had an extraordinarily wide range of interests and left original contributions in numerous areas of mathematics. To most lay people however, he is best known for his creation of what he called "cybernetics." This is a somewhat loosely defined term describing the relation between humans and machines. It becomes more precise in the form of automata theory and control theory. The etymology is from the Greek meaning the management of a vessel. (In fact it was used by the French physicist A.-M. Ampère who used it to refer to the art of government.) The prefix "cyber" has found its way into the realm of computing and the term "cyberspace" is on the lips of even the youngest of children!

Among his fundamental contributions are those in classical analysis, probability theory, and certain aspects of physics as well as physiology. A testimony to his originality is the fact that many basic results bear his name. In addition to his scientific contributions, he wrote on philosophy, religion and he even ventured into the realm of fiction. On the whole, one gets the impression that his physical capacities could not keep up with the endless flow of ideas from his fertile and creative mind.

On the personal side, his character, like his scholarship, spanned a broad spectrum. He has been extolled by some for his generous spirit and tolerance. His relations with others however, could be strained, as when, for example, he resigned from his membership in the National Academy of Sciences. One noted mathematician said, after meeting him for the first time, "he was a poor listener but his self-praise was playful and never offensive."

Not surprisingly, he received many honors in his lifetime and his place in the history of science and mathematics is secure. He was in great demand as a lecturer and indeed it was on one of these tours to Stockholm that he died unexpectedly in 1964.

Editor's Preface

This essay dates from Wiener's youth when he was immersed in philosophy and philosophical speculations. It deals with a fundamental issue of ethics viz. the "Highest Good" or, in the more formal language of ethicists, the "summum bonum." This goal is the essence of human conduct and incorporates concepts of duty, conscience, etc. and continues to be the object of study by philosophers and theologians. Many writers believed that there is an ideal of moral conduct that is fixed and unchangeable, but this point of view is not universally held.

Anticipating the goal of the article, we find that Wiener concludes that there is no *highest good* but that there is a goal which continually changes. He brings to his arguments a sort of mathematical methodology and in the process uses, in simple form, an idea that might legitimately be claimed as a germ of the ideas of his later theory of automata. Before commenting in detail, let us review briefly some of the speculations that had gone before and with which it is certain Wiener was familiar.

It should be pointed out that in dealing with these issues, ethicists and philosophers speculating on these matters inevitably introduce new concepts many of which are inextricably bound up with one another. There is therefore the peril that an argument eventually becomes circular. Nevertheless, the goal of answering the question "what should I do" is sufficiently important to justify the effort.

Plato and Socrates taught that happiness is the supreme and ultimate object of human endeavor and happiness is identical with the highest good. One should not conclude that happiness is the same as pleasure, they hasten to add. Happiness is a state of mind, while pleasure involves a strong component of physical well-being. The pleasure of an agreeable meal is not to be confused with the happiness of feeding a hungry waif. Aristotle extols speculative thought and the contemplation of "honorable and divine subjects." He supplements this by saying that he does not aim at the good that is absolutely supreme but that which is highest for humans in their present condition. We shall see this relativism in Wiener.

Another approach is to assert that the highest good is to be equated with duty or virtue. In particular, Kant claims that the highest good is that which adheres to the moral law whose existence Kant claims can be rationally established. An act is not good because of its consequences but because it conforms to what Kant calls the categorical imperative.

St. Thomas Aquinas taught that man's highest happiness does not consist in pleasure but in action and the highest happiness of man consists in the knowledge of the highest truth, which is God.

Finally we have the traditional views of classical religions. They say simply, that there is a set of revealed laws of behavior that, if adhered to, lead to rewards and if transgressed, lead to punishment here on earth or elsewhere.

It is not possible in a short span to elaborate on all possible views of the highest good. The reader will have noticed the hazard expressed above, that many new concepts have injected themselves, sometimes subtly, and these may be at least as difficult to characterize as the "summum bonum" itself. Indeed it is highly likely that readers will have formulated their own theories and convictions.

To return to Wiener, he grants that there exists a moral righteousness that is not fixed however, but expands with our moral attainments. The fact that one course of action is judged more moral than another does not imply that there is a greatest any more than the existence of an integer greater than any given one implies that there is a greatest integer. Along with Herbert Spencer, Wiener adopts a somewhat Darwinian view. The instincts, behavior, conscience, etc. of members of a species are adapted to maximize the growth and survival of the species. A vital component in the growth is the enhancement of behavior—in ethical terms—to increase the morality, for this necessarily increases the success of the species. Viewed as automata, the members of the species learn from their successes and transmit this learning to subsequent generations; morality thus grows with each stage in the evolution of the species.

Thus the fact that there is no "highest good" does not mean, according to Wiener, that there is no "higher good" to be strived for at each stage in the course of evolutionary change.

We now let Wiener speak for himself.

Many systems of ethics—perhaps most systems of ethics—begin with a theory of the nature of the highest good. It is on the basis of their treatment of this problem that most ethicists can be classed either as hedonists or as perfectionists. Considered as expositions of the *nature* of the *summum bonum*, hedonism and perfectionism are diametrically opposed to one another, but hedonists and perfectionists agree in maintaining that some sort of a *summum bonum* exists, and that the true task of ethics is to inquire into its nature. They believe, that is, that there is one fixed, immutable ideal of all moral conduct, and that the rightness or wrongness of our human actions is to be measured by the degree of completeness with which this ideal is attained.

That there is such a fixed, immutable ideal of morality is usually taken for granted by those ethicists who maintain it without any explicit argument. They usually proceed to treat the question of the nature of this ideal as if it were not distinguishable from the more general question as to the nature of morality *überhaupt*. But one should note that *prima facie*, at any rate, the hypothesis that a highest good exists may well be doubted without involving us in scepticism as to the existence and validity of moral distinctions. It would seem at first sight that the existence of the relation between one object of a moral judgment and a better one does not entail the existence of an object of a moral judgment to which every other such object bears this relation, any more than the existence of the relation between one number and a greater one demands that there should exist some number greater than all other numbers. The notion of a hierarchy of values does not in itself demand that this hierarchy should contain a highest value. The hypothesis that there is no *summum bonum* is, thus, at least worthy of consideration.

Perhaps one of the grounds which leads ethicists to believe in the existence of a *summum bonum* is that they consider that disbelief in its existence, since it demands that we should deny the existence of any single goal of moral conduct, forces us to regard morality as vain and purposeless, and drives us to a pessimistic view of ethics. This is simply false. For to deny that moral conduct has any ultimate ideal is not to say that there are no moral ideals at all. Though we may doubt the existence of an ultimate ideal, we may say that our ideals grow with our attainments, that the better a man becomes, the broader are the vistas of righteousness that open out before him, that to reach the goal that our ideals point out to us does not close our moral development, but simply shows us further and greater goals to strive for and to attain. Surely this is not pessimism! Indeed, it is difficult for me to conceive how even those who believe in the existence of a highest good can regard its attainment as the motivating ideal of human moral conduct, without taking a pessimistic view of morality, for if our highest ideal is capable of definitive attainment, then on its attainment moral progress ceases and morality culminates in becoming a state of repose rather than a way of acting, while if the sole ideal of moral conduct can not be attained, then, in the last analysis, morality is a perpetual failure. Any *partial* attainment of the highest good, if it is to possess any value at all, must possess it by virtue of being the *complete* attainment of something good in itself, though less in value than the *summum bonum*, and hence by virtue of the plurality of possible ideals of moral action. Now it seems to me, at any rate, though, I acknowledge, not to everybody, that any view that either admits the possibility of a cessation of moral progress or the inability of morality to attain its ideals is essentially pessimistic.

Any valid arguments in favor of the existence of a highest good must start, then, not from the mere existence of moral distinctions and moral ideals, but from the particular nature of these distinctions and ideals. We must ask ourselves, what is it that constitutes the goodness of this act and the badness of that one? Is it that this act chimes in with certain abstract moral laws of which we are conscious, and that one does not, or does it mean that a certain nonintellectual faculty within us sets on the one and not on the other the stamp of its approval, or, if neither of these, what does it mean?

There is one thing on which all ethicists will agree: our sole mode of access to the good is through what we call our conscience, though hardly any two ethicists are in precise accord as to just what this "conscience" is. A being without what might in some sense be called a conscience would never arrive at the notions of "right" and "wrong" at all, though he might easily learn to use the *terms* "right" and "wrong" as conventional names for two different sorts of actions. And in fact we mean by saying that a certain act is right or wrong simply that it would be approved or disapproved by some conscience: either by our own, or by the consensus of the consciences of society, or by the conscience of some ideal impartial observer. If there existed no consciences to censure or to approve, there would be no right and wrong nor good and evil.

Moreover, our conscience is not a mere intellectual awareness that certain acts conform to the moral law and others do not. It is one thing to realize in cold blood that a certain act is among those that possess a given formal property—namely, that of satisfying a certain abstract moral law—and quite another thing to feel that *we* ought not to do it. As James says, "When an idea *stings* us in a certain way, makes as it were a certain electric connection with our self, we believe that it *is* a reality. When it stings us in another way, makes another connection with our self, we say, *let it be* a reality. To the word 'is' and to the words 'let it be' there correspond peculiar attitudes of consciousness which it is vain to seek to explain." A mere intellectual awareness that a given act is one of those referred to in a certain manner by the moral law is a different thing from the command, "Let it be performed." Just as one's knowledge of the rules of grammar can only render his speech grammatical when coupled with a *desire* to obey these rules, so one's awareness of the moral law can only issue in action when conjoined with some impulse within which urges one to obey this law. Conscience does not speak in the indicative, but in the imperative mood, and as James tells us, the indicative and the imperative moods represent radically distinct categories of thinking.

Conscience resembles our "feelings" in speaking in the imperative mood. Now, our feelings may be divided more or less sharply into two classes, according as to whether or not they can, if I may put it so, speak in the conditional as well as in the imperative mood. Certain of our feelings, such as hunger, thirst, etc., apply at the present moment to no past objects. We feel no hunger for a dinner we have eaten, nor for one we have missed. On the other hand, those of our feelings that we call prejudices apply to what has gone by as well as to what exists at present. When we read a book of history, we find in it, among other things, many expressions of the author's prejudices in regard to past forms of civilization, society, and government. Similarly our prejudices apply to hypothetical conditions that we know are impossible: for example, many people will feel strongly prejudiced against the state of affairs pictured in Bellamy's "Looking Backwards," or other similar Utopias, though they fully realize that such conditions can never exist. Our mere bodily feelings, however, such as hunger, thirst, etc., can never apply to any but an immediately present object. Our consciences, we can clearly see, are more closely allied to our prejudices than to our bodily impulses in this respect: we make moral judgments about, and have conscientious attitudes toward things that are no more and things that do not exist, have not existed, and never will exist. We conscientiously approve or disapprove of the actions of the characters of history and the characters in a novel. Both our consciences and our prejudices are feelings which may apply to ideal objects.

But this is not the only respect in which conscience and prejudice resemble one another. As a matter of fact, often one person will call a given sentiment of approval of ideal objects a prejudice, whereas another will call it a conscientious sentiment. For example, the Stoic would say that motives of the nature of conscience led him to justify suicide, while we should say that it was a prejudice. And indeed, even if it was a prejudice, there can be no real doubt that the way it felt to him was just like the way our conscience feels to us. It will be clear, if we reflect on this and other similar cases, that the distinction between our conscience and our prejudices can not be one of emotional quality. Feelings that bear every qualitative mark of being conscience are often the most extravagant prejudices, while the feelings which we regard as conscientious will frequently be considered the most outlandish prejudices by people of other races or times. To take an extreme instance, any civilized man would instantly call the feeling of many cannibal tribes that cannibalism is the most respectful way of disposing of the dead a low and detestable prejudice, but I have no doubt that the cannibal would raise up his hands in horror at the prejudices that lead the civilized man both to insult his dead and to waste a valuable source of food by burying the bodies of his dead in the earth. Again, one may regard one of his own emotions now as a conscientious emotion and now as a prejudice; for example, the religious convert is likely to say that the feelings that urged him to observe the ritual of his previous religion were mere prejudices, whereas before he regarded them as due to conscience, while in his attitude to the feelings which urge him to observe the ritual of his present religion, exactly the opposite change may take place.

Nor is it primarily in the nature of the objects of their approval that the difference between conscience and prejudice rests. No mere difference in their objects is sufficient to account for the discrepancy in the obligatory force that we consider them to possess. Moreover, it would really involve a vicious circle to say that conscience differs from prejudice in approving those courses of action that satisfy the moral law, for the sole sources of our knowledge of the moral law are the *dicta* of our conscience. It might be urged that our conscience is the feeling which leads us to approve those courses of action which an impartial observer would approve, but then the question arises, what sort of a person would an impartial observer be? By an "impartial observer" we may mean (1) an observer with no prejudices, but with a conscience, or (2) an observer with neither prejudices nor conscience. To say that our conscience is the feeling that leads us to approve those courses of action that an impartial observer of the first sort—i.e., an observer with no prejudices, but with a conscience—would approve, is clearly circular, whereas an impartial observer of the second kind could neither approve nor disapprove of any sort of action whatsoever. A being with neither prejudices nor conscience would be an utterly unmoral being, to whom good and bad would be alike without value. He would be an unmoved spectator both of the basest crime and the noblest act of benevolence. So, unless some third precise meaning is given to the notion of the "impartial observer" it is of no avail whatsoever in distinguishing prejudice from conscience. In general, we may conclude that the difference between prejudice and conscience is more deeply seated than it would be were it a difference in the nature of their objects alone.

There is one respect, however, in which it is easy to see that our prejudices and our conscience are different, and that is indicated by the fact that we speak of our conscience in the singular number, but of our prejudices in the plural. Those feelings that collectively form our conscience, in general, strengthen the effect of one another upon our action by urging us in the same direction, while our prejudices are in accord, as a rule, neither with one another nor with our conscience. Indeed, when we find various feelings within us at cross-purposes with one another, and can find no method by which to bring them into mutual harmony, we come to the conclusion that perhaps all, and certainly some of them, are not motives of conscience. When I now deem a mere prejudice what I once considered to be a moral emotion, I mean to say that although I once regarded it as harmonizing with the system of my other feelings, I realize in the light of my present knowledge that it conflicts with the most powerful coherent group of feelings within me of such a sort that they may be directed towards ideal objects. Similarly, I approve or condemn the motives of another in accordance as they would respectively cooperate or conflict with that group of feelings capable of ideal objects which, I believe, would be preponderant in me as I am now constituted were I in his place.

But there is no valid *a priori* reason why our entirely different system of feelings capable of ideal direction may not preponderate in another from that which would preponderate in me were I in his place. As a matter of fact, it is clear that when other individuals are put in situations highly analogous to the one in which we find ourselves, their actions, though they bear every external mark of having been motived by conscience, are very different from our own. One might say, it is true, that this discrepancy is due to the fact that they see the situation otherwise than we do, but I doubt whether this difference is always sufficient to account for the difference between their actions and ours. You would find, for example, if you should select a representative group of Englishmen and a representative group of Hindus, each embracing individuals of every stage of intellectual training and development, that there would be a great similarity between the things which one Englishman feels most strongly that he ought to do and the things another Englishman feels most strongly that *he* ought to do, whereas on many points there will not be so marked an agreement between Hindu and Englishman of the same degree of intelligence and breadth of information. Since my conscience at this present moment is simply the most powerful group of feelings within me capable of ideal direction, the word "conscience" may have a different meaning for every one of us.

But if the only way we can arrive at our notions of moral distinctions is through our conscience, and if there is no *a priori* reason why the conscience of one person should agree with that of another, nor even the conscience of a person at one time with the conscience of the same person at another, why is it that morality is not a purely personal matter? What meaning is there in speaking of the objectively good, or even of that which is permanently good with respect to any one individual? The answer to this question can best be reached by looking at mankind from the biological point of view. Biologically considered, the impulses, instincts, and instinctive motives of an animal are means for the preservation of its race. The impulse of the rabbit to run from its enemy, the instinct of the cat to creep up behind its prey and then spring upon it, the instinct of the sheep to follow the leader of its herd, all these are perpetuated from generation to generation because they are essential to the survival of the race, and the progeny of those members of the race which do not possess these impulses, inheriting the deficiency of their ancestors, rapidly become extinct. The instinctive motives of the human race may be looked at in this light. The feelings into which these motives enter must, therefore, possess a certain inherent stability. As such feelings form an important, if not a dominant part of our conscience, they ensure that our conscience at one moment will have much in common with our conscience at another—that our conscience will usually urge us to perform certain sorts of actions. The commands that our conscience normally makes we may call the commands of our *stable conscience*. It is to this conscience that we refer when we say, for example, "At that moment my prejudices overcame my conscience," etc.

Not only does the instinctive element in our motives explain the existence of a stable individual conscience, but it explains the existence of the social conscience. Among animals, some are gregarious in their habits and some are not; some, that is, are physiologically and mentally fitted for cooperative, and some for independent action. Among those that are fitted primarily for independent action, those instincts that concern their behavior towards their fellows play, on the whole, a secondary part in their impulsive life, which consists chiefly in those instincts which urge them to seek and to consume prey, to avoid enemies, etc. But since a race of gregarious animals has, so to speak, staked its whole chance for survival upon its capacity for concerted action, it is absolutely essential that its members should develop instincts and hereditary emotional tendencies which urge them to cooperate with one another in such a manner as to render concerted action possible; that the herd should, so to speak, act like a larger animal, the resultant of the cooperation of its individual members much as the animal body itself is the resultant of myriads of cooperating cells. Now, man is the gregarious animal *par excellence*, and by the process of the survival of the fittest has evolved such instincts and hereditary emotional tendencies as are essential to the concerted action of mankind. These instincts and tendencies form an important part of the framework on which the moral nature of all of us is built. It is the fact that we possess these motives in common that ensures that there shall be a large measure of agreement between the conscience of one man and that of another. In this common basis of innate (though not necessarily congenital) sentiment which all human beings possess in common the so-called social conscience is rooted. And it is what is approved by this basis of innate sentiment, trained and directed by habit and education, that constitutes the objective good.

It is clear, then, that the social conscience and the objective good are not absolutely fixed. The impulses and tendencies common to all members of the human race have been modified, are being modified, and will be still further modified by the gradual process of organic evolution, which remodels our instincts, and the rapid process of social evolution, that remodels our habits. And among those impulses that seem to be the most subject to evolutionary change are those which speak with a "should" instead of with a "must." The primeval man in all probability felt the commands of hunger and thirst just as we of today do, but he would be at a complete loss to understand our moral scruples, and, I believe, we should be equally at a loss to understand his. The impulses capable of ideal direction are among the latest to appear in the scale of evolution, and it is extremely doubtful, indeed, whether they are to be found at all except in man. They vary enormously from race to race and from age to age. They are influenced by training far more than our cruder bodily impulses. And so we are forced to say that even the objective good is by no means unchanging.

What is more, it is difficult to determine what, strictly speaking, is the objective good of humanity at any one moment. The human race is subdivided into many not strictly definable parts, each of which differs much from every other in its traditions, and considerably in its hereditary equipment of instincts. Each one of these has a more or less distinct objective good of its own, and these objective goods often directly antagonize one another. To the members of one race, a certain course of action may not only seem, but be objectively right, and to the members of another, objectively wrong. Yet, though the various objective ethical standards of different races or peoples do not harmonize in the valuation they give to particular courses of action, we have no way of placing ourselves over and above the differences of these standards, and calling one of them absolutely correct, and the rest absolutely wrong—just as according to the theory of relativity, though the numerical magnitude of the velocity of a body is dependent, among other things, on the *Bezugssystem* to which we refer it, there is no definitive *Bezugssystem* which can be picked out from all the others, and called the right one. And just as in physics, we choose as our *Bezugssystem* that in which some body that interests us—such as some point on the earth's surface, or the center of gravity of the earth, or the center of gravity of the solar system—is approximately at rest, so in ethics we treat that objective good as if it were definitive in which our permanent conscience, or that of our family, our class, or our nation is justified. A person of another race may use a different ethical *Bezugssystem* just as a person on another planet would use a different physical *Bezugssystem*. The ethical standard that it is natural for us to use, though it is not *a priori* superior to that which is natural to another, may drive me to actions which conflict with his. Two races may come into a war in which each is, from its own standpoint, absolutely right, and from that of the other, absolutely wrong. In fact, it is only the instinctive feeling within which urges us to respect the consciences and prejudices of other human beings which ever prevents us from overriding the consciences and prejudices of others when they conflict with our own. And this feeling, though we unquestionably all possess it, is frequently overwhelmed by the force of the rest of our conscientious feelings opposing its application in a given instance.

Again, even within the race and the nation, the consciences of certain individuals run counter to the social conscience. There are so-called moral imbeciles, whose conscientious feelings are aborted, and morally insane people, whose consciences are malformed, and persons of these two classes often commit actions which the social conscience disapproves. Now, since we have said that a person's private good is what his conscience approves, and since the consciences of these individuals, such as they are, either approve or do not disapprove of their actions, it might be asked, by what right does the social conscience demand that they should be punished, or at any rate restricted in their opportunities for opposing and injuring society? Have we not said that there is no impartial observer of two consciences, and hence no impartial way of comparing the criminal's conscience with that of society? Then how is the social conscience any better than the conscience which it condemns? The answer here is clear: the social conscience need finally consider only one good—the good to which it itself urges society. Except in so far as this good happens *de facto* to involve the respecting of the prejudices of another, even though they lead to antisocial acts, the social conscience may utterly disregard that of the criminal—and similarly the criminal that of society. *If the conflict between the criminal and society, or that between two peoples having moral standards which irreconcilably antagonize one another, can not be settled by altering the outlook of those who are on one side of the controversy so that their consciences are changed in such a manner as no longer to conflict with those of the members of the other side, and if no third view of the situation can be developed on which the consciences of the disputants on both sides will agree, the conflict can be settled, if at all, only by the suppression by brute force of the disputant or disputants on one side.*

This view may seem a return to the position of Hobbes, since we regard force as a final arbiter of moral disputes, but it differs radically from Hobbes's view in that it does not consider man fundamentally selfish, nor morality as based upon a purely external contract between the naturally discordant members of a nation. As Hobbes claims, objective morality is of the nature of a compromise, but this compromise between the actions demanded by the feelings of the various members of society is only possible because of the immense common ground of sentiment that all normal human beings possess in common. Among the individual feelings of a human being, which are the bases of all morality, are other-regarding as well as self-regarding feelings.

Objective morality is, then, the end product of a double conflict: the conflict among the various feelings capable of ideal objects within each individual human being, and the conflict between the private consciences of the various members of mankind. And these struggles were not finished in the distant past, but are going on at the present moment. Objective morality is changing continuously, and this change is not merely in what we *call* the objective good, but in the objective good itself, if the phrase "the objective good" is to have any meaning for us. What *is* good one moment may be bad when looked back on from the next. The ideal of morality today may be attained, or displaced from its position as an ideal. There is no single universal ideal of morality the same for all time and all humanity; morality is human and mutable. Though it may be that our morality approaches some limit as it develops, such a limit is recognized as absolutely good from no stage in the process of moral development, and, indeed, *is* not absolutely good at any stage of that process. There is no highest good.

Sources of the Articles

Mary L. Cartwright, "Mathematics and Thinking Mathematically," *Amer. Math. Monthly* 77 (1970) 20–28.

Henri Poincaré, "Mathematical Invention" (L'invention mathématique), *Enseignement Math.* 10 (1908) 357–371. Translated from the French by the editor.

Jacques Hadamard, "Thoughts on the Heuristic Method" (Réflexions sur la méthode heuristique), *Revue Gén. Sci.* 16 (1905). Reprinted in *Œuvres de Jacques Hadamard*, vol. 4, Editions du Centre National de la Recherche Scientifique, Paris, 1968, pp. 2127–44. Translated from the French by the editor.

G. H. Hardy, "Mathematical Proof," *Mind* 38 (1929) 1–25. (Abridged) Reprinted in *Collected Papers of G. H. Hardy,* vol. VII, Clarendon Press, Oxford, 1966–79, pp. 581–606.

Hermann Weyl, "The Unity of Knowledge," *H. Weyl, Gesammelte Abhandlungen,* vol. 4, Springer, Berlin-Heidelberg, 1968, pp. 623–630.

Marston Morse, "Mathematics and the Arts," *Bulletin of the Atomic Scientists* 15, no. 2 (1959) 55–59. Reprinted from *The Yale Review* 40 (1951) 604–612. Reprinted in *Collected Papers/Marston Morse,* vol. 4, World Scientific, Singapore, 1987, pp. 2272–76.

George D. Birkhoff, "Intuition, Reason and Faith in Science," *Science* 88, no. 2296 (December 30, 1938) 601–9. Reprinted in *George David Birkhoff/Collected Mathematical Papers,* vol. 3, American Mathematical Society, New York, 1950, pp. 652–59.

David Hilbert, "Logic and the Understanding of Nature" (Naturerkennen und Logik), *Naturwissenschaften* (1930) 959–63. Reprinted in *David Hilbert/Gesammelte Abhandlungen*, Band III, Springer, Berlin, 1970, pp. 378–87. Translated from the German by the editor.

Raymond L. Wilder, "The Cultural Basis of Mathematics," *Proc. International Congress of Mathematicians,* vol. 1, American Mathematical Society, Providence, 1950, pp. 258–71.

James Joseph Sylvester, "Presidential Address to the British Association," *The Collected Mathematical Papers of James Joseph Sylvester*, vol. 2, University Press, Cambridge, 1908, pp. 650–61.

John von Neumann, "The Mathematician" in Heywood, Robert B., ed., *The Works of the Mind*, University of Chicago Press, Chicago, 1966, pp. 180–96. Reprinted in *John von Neumann, Collected Works*, vol. 1, Pergamon, New York, 1963, pp. 1–9.

André Lichnerowicz, "The Community of Scholars" (La communauté des savants), *Enseignement Math.* 1, sér. 2 (1955) 30–43. Translated from the French by the editor.

André Weil, "History of Mathematics: Why and How," *Proc. International Congress of Mathematicians, Helsinki, 1978,* vol. 1, Academia Scientiarum Fennica, Helsinki, 1980, pp. 227–36. Reprinted in *Œuvres Scientifiques/Collected Papers*, vol. III, Springer, New York, 1979, pp. 434–42.

Paul Lévy, "Does God Exist?" (Dieu existe-t-il?), *Quelques aspects de la pensée d'un mathématicien*, A. Blanchard, Paris, 1970, pp. 181–91. Translated from the French by the editor.

Wilhelm Maak, "Goethe and Mathematics" (Goethe und die Mathematik), *Semesterbericht Göttingen, Math.-Physicalische Klasse*, 1949, 138–49. Translated from the German by the editor.

Francesco Severi, "Leonardo and Mathematics" (Leonardo e la matematica), *Scientia*, 58 (1953) 41–44. Translated from the Italian by the editor.

Norbert Wiener, "The Highest Good," *J. Phil. Psych. and Sci. Method* 11, no. 19 (1914) 512–20. Reprinted in *Norbert Wiener/Collected Works with Commentaries*, vol. 4, P. Masani (ed.), MIT Press, Cambridge, 1985, pp. 41–49.

Index